Corrosion Resistance

Corrosion Resistance

Editor

Albert Brose

scitus academics

Corrosion Resistance
Edited by **Albert Brose**

ISBN: 978-1-68117-305-4
Library of Congress Control Number: 2016937613

SCITUS Academics LLC,
www.scitusacademics.com
Box No. 4766, 616 Corporate Way,
Suite 2, Valley Cottage,
NY 10989

Printed in United States of America on Acid Free Paper

Preface

Corrosion is a natural process that converts a refined metal into a more stable form like its oxide, hydroxide or sulfide. It is the gradual destruction of materials by chemical environmental reactions. Corrosion engineering is the area of corrosion control. Depending on the rate of corrosion resistance, different materials intrinsically have this property. Some other methods, such as painting, galvanizing and combining these methods with coating, can also be used to withstand corrosion. Corrosion is essentially the process through which a material is oxidized and loses electrons from the environment. Corrosion resistance is thus the ability to withstand the binding energy of metal and the deterioration and chemical disintegration during surface exposure to such an environment. There are no corrosion-resistant materials in all surroundings. Protective coatings are the most widely used corrosion control technique. Protective coating materials are essentially corrosion resistant and can separate corrosion-prone surfaces. While cathodic protection interferes with the natural effect of electrochemical corrosive cells. Cathodic protection can be used efficiently to resist corrosion in water or soil-exposed surfaces. Corrosion resistance is an important factor for the choice of corrosion control material.

This book covers state-of-the-art technology, development and research in corrosion studies in a variety of research and application fields. The use of corrosion resistance materials also gives readers in different fields. In addition to traditional corrosion studies, the book also includes chapters on energy, fuel cells, daily materials, green materials and the semiconductor industry.

Table of Contents

CHAPTER 1

Corrosion Resistance Properties of Aluminum Coating Applied by Arc Thermal Metal Spray in SAE J2334 Solution with Exposure Periods

Han-Seung Lee [1], Jitendra Kumar Singh [1,2], Mohamed A. Ismail [3,*] and Chinmoy Bhattacharya [2]

[1] Department of Architectural Engineering, Hanyang University, 1271 Sa 3-dong, Sangrok-gu, Ansan 426-791, Korea

[2] Department of Chemistry, Indian Institute of Engineering Science and Technology, Shibpur, Howrah 711103, India

[3] Department of Civil and Construction Engineering, Faculty of Engineering and Science, Curtin University Sarawak, CDT 250, 98009 Miri, Malaysia

ABSTRACT

Arc thermal metal spray coating provides excellent corrosion, erosion and wear resistance to steel substrates. This paper incorporates some results of aluminum coating applied by this method on plain carbon steel. Thereafter, coated panels were exposed to an environment known to form stable corrosion products with aluminum. The coated panels were immersed in Society of Automotive Engineers (SAE) J2334 for different periods of time. This solution consists of an aqueous solution of NaCl, CaCl2 and NaHCO3. Various electrochemical techniques, i.e., corrosion potential-time, electrochemical impedance spectroscopy (EIS) and the potentiodynamic were used to determine the performance of stimulants in improving the properties of the coating. EIS studies revealed the kinetics and mechanism of corrosion and potentiodynamic attributed the formation of a passive film, which stifles the penetration of aggressive ions towards the substrate. The corrosion products that formed on the coating surface, identified using Raman spectroscopy, were Dawsonite (NaAlCO3(OH)2) and Al(OH)3. These compounds of aluminum are very sparingly soluble in aqueous solution and protect the substrate from pitting and uniform corrosion. The morphology and composition of corrosion products determined by scanning electron microscopy and energy dispersive X-ray analyses indicated that the environment plays a decisive role in improving the corrosion resistance of aluminum coating.

KEYWORDS

passive film; arc thermal metal spray; electrochemical impedance spectroscopy; Raman spectroscopy; SEM; X-ray diffraction

1. INTRODUCTION

Steel is being used everywhere in world due to its broad application. Steel has inherent mechanical properties to a proper extent, but there are some problems related to corrosion, especially in aqueous, dry and atmospheric conditions. The atmospheric corrosion of steel is a serious problem and different types of organic and inorganic coatings have been developed to extend the service life of structures made from steel. Besides coatings on structural steel, stainless steel is being used in infrastructure but is not economical. The coating of steel with aluminum metal is found to provide protection in industrial and marine environments. Different techniques for the application of aluminum coating such as hot dip, aluminizing, high velocity oxy fuel, plasma spray and arc thermal spray methods, have been developed. Among these techniques, arc thermal spray methods are convenient, economical and logistically feasible [1,2]. To protect bridges, strobes, pipelines and other large steel structures for long-term protection in industrially polluted, saline and open atmospheres, the use of Al coating with the thermal spray process had been reported about a century ago [3]. The application of coatings with the arc thermal process is very easy and convenient. In this process, prior to the application of the coating, the target surface should be proper cleaned and sandblasted because a good anchoring for the coating is required. This instrument gun is portable and a power supply is needed for the arcing of wires. Before applying the coating, it must be ensured that there is a proper power supply at the application location and a fuel tank necessary for the flame spray. In the arc thermal metals spray process, twin wires are used for coating and an electric arc melts the tips of these wires. After arcing, the wires melt and atomized by hot air which impinges on the target (substrate). The melted metal droplets adhere onto the substrates and are cooled at room temperature, resulting in the formation of a coating [4]. Due to the high speed of spraying and the sudden cooling of melted metal droplets spread at the substrate, some invisible porosity in the deposited coating and shrinkages develops [1,5,6,7,8].

To fill the pores or defects in coatings, various methods have been suggested. The most popular method is metal-polymeric dual coating. The application of polymeric coating is better for shorter durations of exposure with regard to corrosion protection. In the case of longer durations, it causes cracking or premature distresses in the coating. It is caused due to a difference in the thermal expansion coefficients of the metal-polymer. Li et al. (2006) and Muhamad et al. (2013) have carried out research into the protection of steel by means of Al coating through an arc thermal spray process with a filler; they reported enhanced corrosion resistance properties for this coating after eight years of exposure in a coastal environment [9,10].

A search of the literature reveals that very limited information is available on the use of fillers or other techniques to block the porosity of arc thermal sprayed aluminum coating. To enhance the properties of coatings, researchers are using different pseudo alloys and rare earth metals for corrosion resistance properties in aggressive environments [11,12]. Due to the anodic nature of Al metal, it sacrificially protects the steel surface. During the corrosion process of the Al coating, its corrosion products are deposited on the coating surface. These corrosion products fill the pores of the coating and enhances the corrosion resistance performance of the Al coating by arc thermal spray process. The most popular test environment to assess the performance of the coatings is 3.5 wt. % sodium chloride. The corrosion performance of Al coatings in coastal and open industrially polluted environments showed the enhanced corrosion resistance properties for longer duration of exposure due to the sacrificial nature of Al metal [13] and the formation of naturally occurring protective corrosion products on the coating surface [14]. These protective corrosion products block the pores of the coating and increase the corrosion resistance properties of the coating. For studies on atmospheric corrosion, the coating should be exposed for longer durations to assess its corrosion resistance properties [15].

A search of the literature also shows that there is possibly no such research being carried out by anyone to study the performance of Al coatings in the laboratory in the presence of stimulants. Generally, stimulants are induced in nature for the corrosion of metals and alloys. In the atmosphere, there are many stimulants present, i.e., carbonates, chlorides, sulphates, nitrates, etc. The atmospheric corrosion for coated samples requires a significant amount of time to determine their performance and depends on the aggressiveness of the environment [15]. However, this study simulates the aggressive atmospheric environment to form naturally occurring corrosion products to block the pores of the coating in immersion conditions. We allowed the deposited coating to age in an environment having a lower concentration of sodium chloride, sodium bicarbonate and calcium chloride. The idea was to develop sparingly soluble corrosion products under immersion conditions and block the porosity in the coating. The objectives of this study are as follows:

To evaluate the performance of Al coatings in immersion conditions of SAE J2334 solution;

To determine the kinetics and mechanism of corrosion process for Al coatings applied by arc thermal metals spray in the presence of stimulants, i.e., carbonate and chloride ions;

To study the role of stimulants on the formation of corrosion products on the surface of Al coatings in SAE J2334 solution;

To study the nature of corrosion products formed on Al coating surface in presence of stimulants, i.e., carbonate and chloride ions.

2. MATERIALS AND METHODS

2.1. Process of Coating

An arc thermal metal spray coating with commercially pure Al (99.95 wt. %) wires of 1.6 mm diameter was applied to a sandblasted mild steel substrate. The chemical composition of the mild steel is shown in Table 1. The dimension of the samples was 80 mm × 60 mm × 1.0 mm. The thickness of the coating was measured using a non-destructive technique with Elcometer 456 at different places, and was approximately 100 μm (±5 μm). The details of the spraying parameters are shown in Table 2 [16,17,18].

Table 1. Chemical composition of steel (wt. %).

Chemical Composition (wt. %)								
C	Si	Mn	P	S	Cu	Cr	Ni	Fe
0.240	0.260	0.950	0.016	0.008	0.020	0.040	0.030	balance

Table 2. Detail parameters used in the high frequency arc thermal metal spray process [16,17,18].

Parameters (unit)	Quantitative Value
Air Pressure (bar)	4–6
Spraying distance (cm)	15–25
Spraying Voltage (V)	30
Spraying Current (A)	200

In this method, Al metal is being melted at an arc point through a circular slit by hot air stream in which metal can be diffused and cooled at room temperature (Figure 1). The detail of the spraying parameters was maintained on the basis of earlier work done by different researchers [16,17,18]. The diffused layer of coating was formed on the steel substrate. During the solidification and diffusion of metals towards the steel substrate, some pores are being formed on the coating. The pores are different sizes and contribute to getting moisture, oxygen and other aggressive ions from the atmosphere.

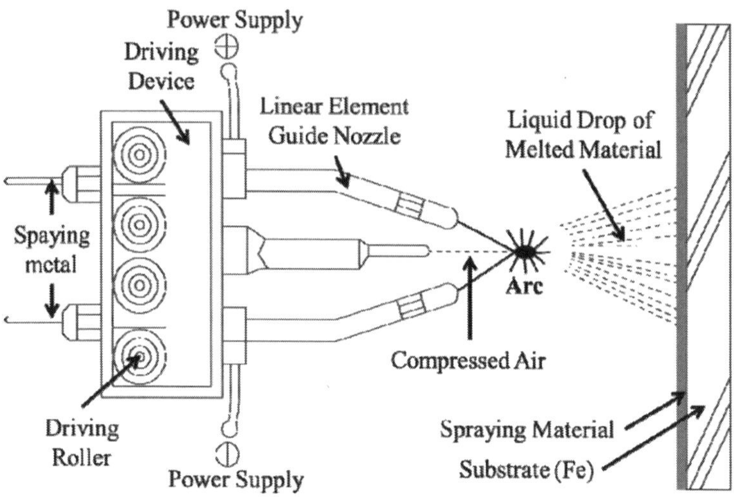

Figure 1. Schematic diagram of the arc thermal metal spay process.

After the application of the coating on the steel substrate, the adhesion of the coating strength was measured based on to ASTM D4541 [19].

2.2. Electrochemical Studies

The electrochemical studies were performed with the SAE J2334 (Society of Automotive Engineers) solution [20,21,22,23,24]. It consists of 0.5 wt. % NaCl, 0.1 wt. % CaCl2 and 0.075 wt. % NaHCO3. The Al coating is directly immersed in the SAE J2334 solution and it is likely identical to atmospheric weathering conditions, which contain chloride and carbonate ions in open industrially polluted environments. The chemicals used were of analytical grade. The pH of the SAE J2334 solution was measured at room temperature and was 9.17. These chemicals were dissolved in double distilled water. Prior to the start of the experiments, the samples were exposed to the solution and stabilized the potential with potentiostat. These studies were performed by three electrode systems where the coated sample worked as a working electrode (WE), platinum wire as a counter electrode (CE) and silver-silver chloride as a reference electrode (RE). The electrochemical setup is shown in Figure 2. The sample holder area of working electrode was 0.78 cm2 and it was fixed for all samples.

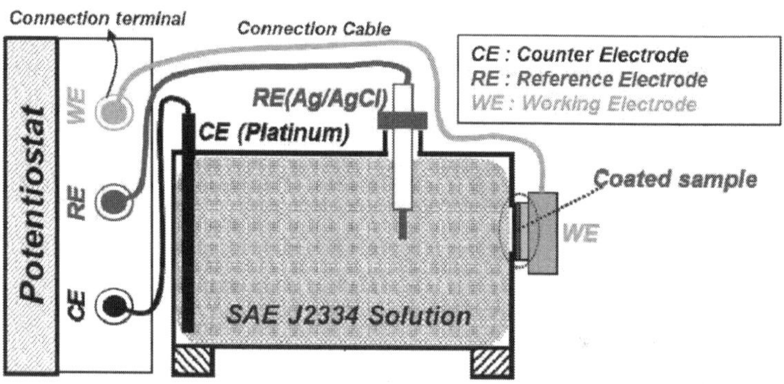

Figure 2. Schematic diagram of the electrochemical setup.

The electrochemical impedance spectroscopy (EIS) studies were carried out by changing the frequency of 10 mV sinusoidal voltage from 100 kHz to 0.01 Hz. The DC polarization studies were performed at a 1 mV/s scan rate from −0.4 V to +0.8 V Vs open circuit potential [11,25]. The potentiostat was VersaSTAT (Princeton Applied Research, Oak Ridge, TN, USA) and data analysis was carried out using Metrohm Autolab Nova 1.10 software (Ionenstrasse, Herisau, Switzerland) by fitting the experimental data in the constant phase element (CPE) model. All electrochemical studies were carried out at room temperature (27 ± 1 °C).

2.3. Characterization of Coating and Corrosion Products

The morphology of coating and corrosion products were determined by Scanning Electron Microscope (SEM, Philips XL 30, North Billerica, MA, USA) operated at 15 kV, equipped with an Energy Dispersive X-ray Spectroscopy (EDS) tool for elemental analysis. Prior to taking the image of the SEM, samples were coated with platinum to increase the conductivity and avoid the charging effect of samples. The X-ray diffraction (XRD, Philips X'Pert-MPD, EA Almelo, The Netherland) studies of the coating was performed by using Cu Kα radiation (λ = 1.54059 Å) generated at 40 kV and 100 mA. The Raman spectroscopy (Renishaw RM 1000, Mercury Dr., Champaign, IL, USA) of corrosion products was carried out by using a He-Ne laser beam of 632.8 nm wavelength. The power of the laser was kept at 10 mW to avoid the transformation of corrosion products due to the heating effect. The collection time was 10 s and the ranges of Raman shift between 200 cm−1 and 3200 cm−1. The locations of the specimens to be studied were focused through an Olympus microscope at a magnification of 20 (Mercury Dr., Champaign, IL, USA). The sample holder had a motorized platform with Jokey (Mercury Dr., Champaign, IL, USA) in order to have fine focusing and mapping at a desired location. Prior to the analysis of the samples, the instrument was calibrated by using pure silicon at the peaks of 520 cm−1.

3. RESULTS AND DISCUSSION

3.1. Characterization of the Arc Thermal Sprayed Coatings

The coating was characterized by SEM and XRD. The top surface and cross section morphology of the coating by the SEM are shown in Figure 3. This coating had many defects as shown by the arrow marks (Figure 3a,b), which is a well-known feature for such coatings [6,8]. The melted metal particles continuously impinged on the base metals and formed layer upon layer of plate-like microstructures [26]. With the high velocity deposition of the coating and after rapid cooling at room temperature, the air diffused from the coating surface resulted in the formation of a splashed zone or pores/defects on the coating surface [27,28]. These splash zones contributed to getting moisture and other aggressive ions from the atmosphere [29,30]. A chemical analysis of the coating is shown in the EDS spectra (Figure 3c). On the surface of coating, 2.20 wt. % oxygen was present; this oxygen might have come during the spraying of the coating at high temperature and in-flight particles oxidation from the atmosphere [31].

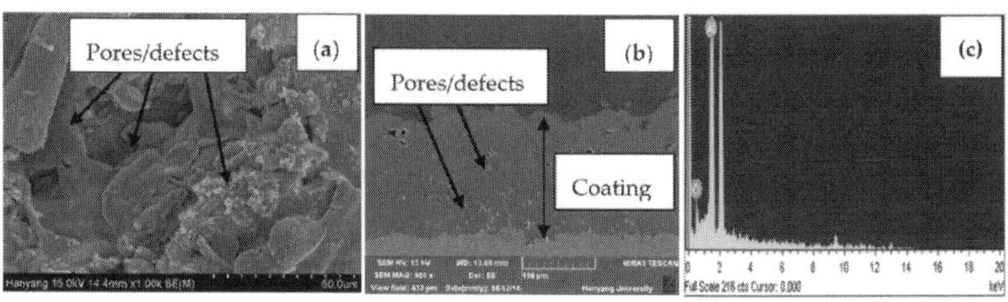

Figure 3. SEM images (a) top surface, (b) cross section and (c) EDS of the Al coating by the arc thermal metal spray process.

The XRD of the coating is shown in Figure 4 and, from this figure, it can be seen that after the spraying of Al using the arc thermal metal spray process, pure Al is present. There is no other phase present on the coating surface besides Al, because the oxygen concentration is very low and the oxide phase which might be present is beyond the limit of XRD detection.

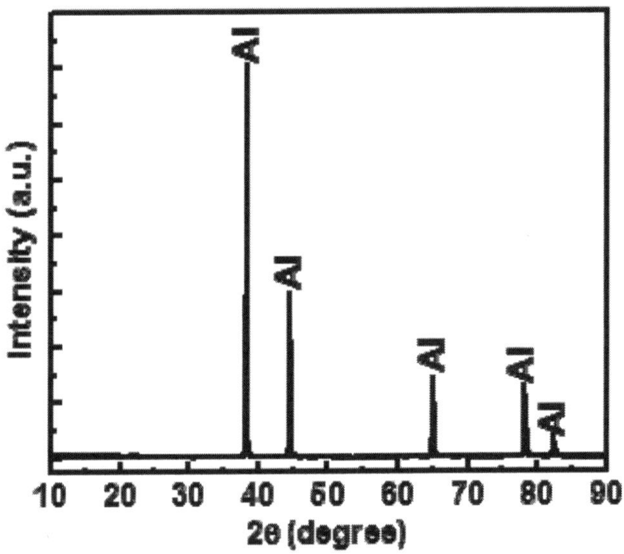

Figure 4. The XRD of aluminum coating using the arc thermal metal spray process.

The bond strength of the coating was measured and is presented in Table 3. The coating was applied on a sandblasted surface for proper adhesion to the steel substrate. In this process, 40 mm × 40 mm of selected dimensions of the coating substrate were used for the adhesion test. The bond strength measurements were carried out for four samples, and the average was taken for all samples. The average coating strength was 4.86 MPa and the adhesion also depends on the feedstock value [32,33]. The bond adhesion of applied coating was low and is due to more selection of the coating surface area.

Table 3. Bond adhesion test results of the Al coating using the arc thermal metal spray process.

Sample Number	Bond Strength (MPa)
1	4.81
2	4.97
3	4.83
4	4.82
Average	4.86

3.2. Electrochemical Studies of Coating

3.2.1. Potentiodynamic Studies

The corrosion resistance properties of the coating were evaluated in the SAE J2334 solution. Prior to this, the coating was exposed in solution for 1 h to stabilize its potential. To ensure the stabilized potential, this experiment was carried out with a potentiostat. Thereafter, potentiodynamic studies were carried out. It can be seen from Figure 5 that the coated substrate exhibited more resistance to corrosion than bare steel, and that the corrosion potential (Ecorr) for the coated substrate is more active. Due to more active Ecorr of the coating in the SAE J2334 solution, it was found that Al metal provided cathodic protection to the steel substrate [13]. The electrochemical data are shown in Table 4, which were extracted from fitting these curves in the Tafel region. The Ecorr values are −0.689V and −0.717V, and the Icorr (corrosion current density) values are 15.22 μA/cm2 and 12.96 μA/cm2 for bare steel and Al coating, respectively. Once the Icorr is determined, the corrosion rate (CR) can be calculated using Faraday's law and putting these value in the following equation [34]:

$$\text{Corrosionrate}(\mu\text{m/year}) = \frac{3.27 I_{\text{corr}} \cdot \text{E.W.}}{d} \quad (1)$$

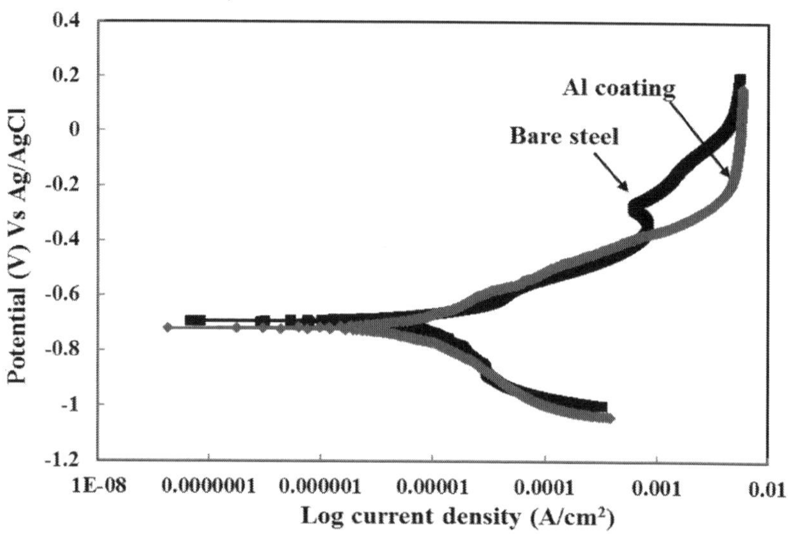

Figure 5. Potentiodynamic studies of samples in the SAE J2334 solution after 1 h of exposure.

The corrosion rate in Equation (1) is expressed in micrometers per year (μm/year), corrosion current density (Icorr) in μA/cm2. The Icorr was obtained by dividing total surface area of the working electrode in the corrosion current, E.W. represents the equivalent weight (g/mole) and d is the density (g/cm3).

Table 4. Electrochemical parameters extracted from potentiodynamic polarization curves in the SAE J2334 solution after 1 h of exposure.

Sample ID	Electrochemical Parameters			
	E_{corr} (V) *vs.* Ag/AgCl	R_p ($\Omega\cdot cm^2$)	I_{corr} ($\mu A\cdot cm^{-2}$)	CR (μm/year)
Bare steel	−0.689	4155.0	15.22	176.82
Al coating	−0.717	5634.0	12.96	141.26

From Table 4 it can be seen that the corrosion rate (CR) for bare steel is higher than Al coated substrate in the SAE J2334 solution. The corrosion current density is slightly lower for Al coated substrate than for bare steel, which indicates the lower corrosion rate of coated substrate. The polarization resistance, (Rp) 5634 Ω·cm2 for Al coating and 4155 Ω·cm2 for bare steel, are calculated after fitting the curve in the Tafel region. These results indicate that the bare steel is susceptible to this environment, but a coated substrate provides protection. The bare steel attained another breakdown potential (approximately −0.270 V) at a more anodic side due to the accumulation of corrosion products on the bare steel surface, which were formed during the initial exposure. Coated steel exhibited passive behavior during the anodic polarization, while bare steel exhibited more pitting and it did not form a passive film (Figure 5). From this study, it can be said that the coated substrate has the passivating tendency in the SAE J2334 solution. Higher Rp and lower Icorr correspond to a high degree of protection provided by Al coating, while bare steel exhibited the reverse. These results suggested that the bare steel is more susceptible to corrosion in the presence of such stimulants.

3.2.2. Potential-Time Studies of Al Coating in SAE J2334 Solution.

Coatings were exposed to the SAE J2334 solution and the corrosion potential was measured with the exposure periods (Figure 6). The active potential range was from −0.616 V to −0.910 V for 1 h to 72 h (inset Figure 6), respectively. The shifting of potential may be due to the coating surface's connected porosity, which makes the steel as cathode and gives a mixed potential value. For the initial exposure of the coating in the solution, it activates the surface due to the presence of activating agents such as chloride and carbonates in the solution, while more defects or active sites on the coating surface enhances the dissolution [35]. At the time of exposure, it was passivating the surface, but not actively. Therefore, at time of the initial exposure, the passive film was weak on the surface but once there was a significant amount of reaction between the solution and the coating surface, the surface became more protective. This resulted in a shifting of potential towards the nobler side. After 72 h of exposure, the potential ennobling up to 384 h (−0.810 V) indicated the passive behavior of the coating and determined the coating characteristics in this solution. The shifting of the corrosion potential towards the nobler side is attributed to the formation of passive film on the coating/solution interface [36,37,38]. The formation of passive film enables the higher degree of protection against the coating in the SAE J2334 solution. Thus, the Al coating exposed in simulated accelerated atmosphere conditions provided improved protection against deterioration due to the formation of passive film.

3.2.3. Electrochemical Impedance Spectroscopy (EIS) Evaluation of Coating in SAE J2334 Solution

EIS is a very useful technique for the determination of corrosion characteristics at the solution/coating interface. The EIS evaluation of Al coatings was performed in the SAE J2334 solution at the open circuit potential with exposure periods. The EIS graphs are shown in Figure 7 and Figure 8. The Al coating applied using the arc thermal metal spray methods provided protection to the steel substrate and can be described by EIS studies.

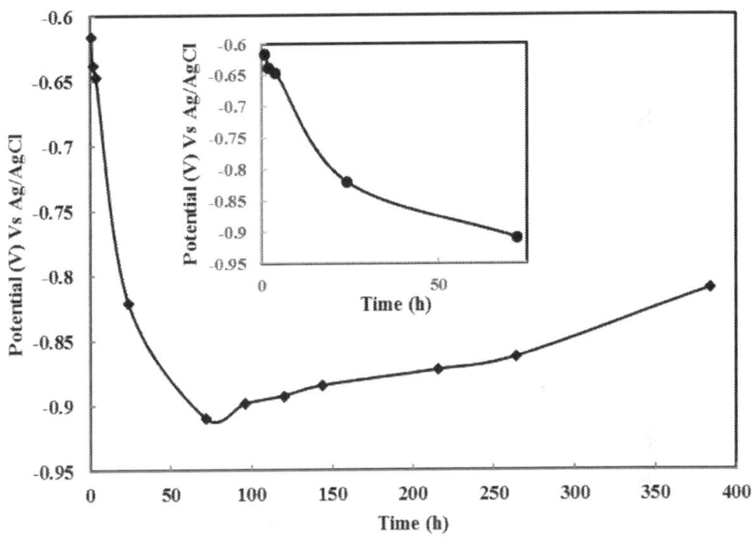

Figure 6. Corrosion potential-time plots of Al coated steel in SAE J2334 solution with exposure periods.

From Figure 7 and Figure 8, it can be observed that the polarization resistance of the coating is continuously increased with exposure periods. From the Nyquist plots (Figure 7a and Figure 8a), the dimension of the semicircle loop has increased considerably with exposure periods, and this is attributed to the formation of a double layer capacitance and a lower corrosion rate [39,40]. As the exposure period increases, the passivating behavior of the coating increases, resulting in an increase in polarization resistance. The substantial increase in magnitude of the semi-circle loop results in the decrease of the active surface of the coating and the formation of a protective layer simultaneously. This phenomenon is attributed to the deposition of corrosion products, which block the defects due to a reaction between the coating surface and the ions (chloride and carbonate) available in the solution [41]. After 384 h (Figure 8a) of exposure, the enlargement of the semi-circle loop is the highest, which means the formation of passive film is uniform, dense and very protective, and that this film exhibited more resistance to electrochemical reaction.

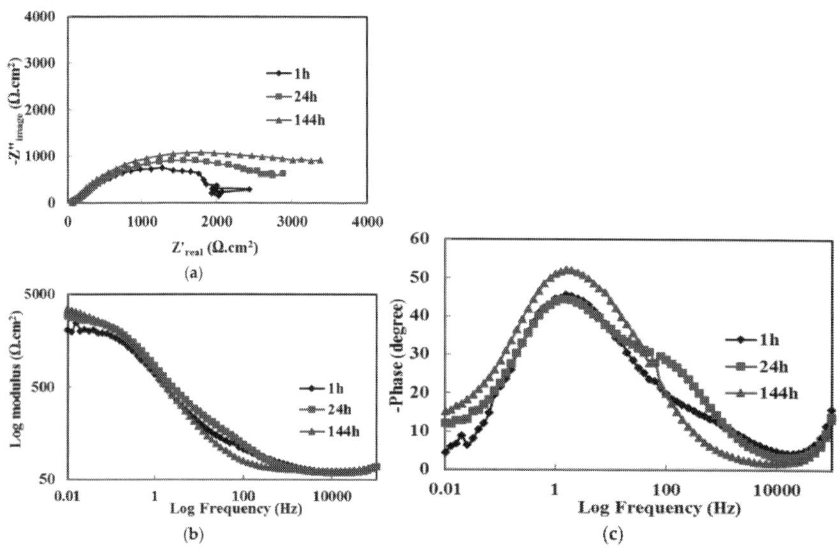

Figure 7. Electrochemical impedance studies of Al coating in SAE J2334 solution with exposure periods (a) Nyquist, (b) Bode impedance-frequency, (c) Bode phase-frequency plots.

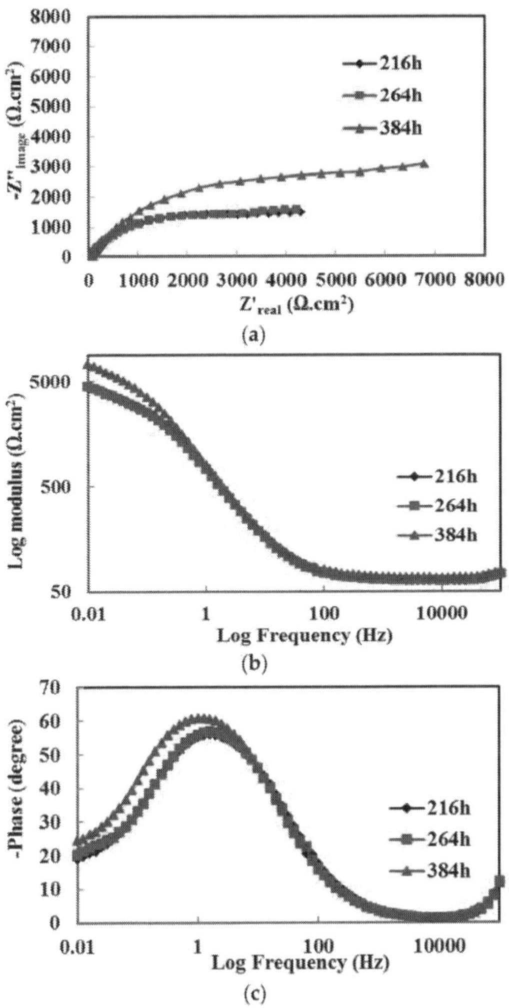

Figure 8. Electrochemical impedance studies of Al coating in SAE J2334 solution with exposure periods (a) Nyquist, (b) Bode impedance-frequency, (c) Bode phase-frequency plots.

Figure 7b and Figure 8b show the typical impedance-frequency Bode plots for coated samples at comparative modes for different exposure periods, and the impedance at the lowest studied frequency (0.01 Hz) that are increased with the progression of immersion periods of the coating in the solution. The gradual increment of the impedance value with the exposure periods exhibited the formation of a protective passive oxide film on the coating surface. The increment in impedance values are attributed to the growth of passive film thickness and the increase in film resistance itself towards the penetration of aggressive ions through the film [42]. This finding also corroborates the results of the Nyquist plots. The increments of the impedance values are more significant for 384 h of exposure than for other exposure periods.

Figure 7c and Figure 8c of frequency-phase Bode plots show that the maximum phase angle for 1 h and 384 h of exposure is −40° and −60°, respectively, and is shifted towards the lower frequency (6 Hz–1 Hz for 1 h and 4 Hz–0.3 Hz for 384 h). The shifting of the phase angle at the lower frequency is attributed to the formation of double layer capacitance and the reduction in the anodic surface area of the coating [43,44,45]. This phenomenon indicates that the surface was covered with protective and thick passive layers on the coating surface. The corrosion characteristics of the Al coating in the SAE J2334 solution exhibited more passive behavior than in the 3.5 wt. % NaCl solution. This phenomenon is attributed to the presence of bicarbonate and chloride ions in the solution, which is responsible for formation of protective passive film of Al containing carbonate and hydroxides. This mechanism will be described in a subsequent paragraph.

After fitting the impedance data in an appropriate electrical equivalent circuit (Figure 9), this model was also reported by other researchers [46,47,48]. The impedance data are best fitted in the CPE model because coatings have inhomogeneity, defects and rough surfaces, which allow the solution to penetrate across the coating surface [49,50,51]. The electrochemical parameters were calculated and the data is presented in Table 5. In Figure 9, Rs, CPEc and Rpore are representing solution resistance, coating capacitance and polarization resistance, respectively. From Table 5 it is observed that the Rpore of the Al coating has continuously increased with exposure periods in the SAE J2334 solution and is exhibited due to the formation of more protective passive layers on the coating surface. Since Rpore increases, the Yo (admittance) should be decreased and it is confirmed by the calculated values of electrical parameters. Initially, the coating surface was rough so that the Yo was high and indicated higher double layer capacitance [52], but as exposure periods are increased, this value became less. This result indicates that the passive film became non-porous and protective. This observation is exhibited due to the formation of strong passive film on the coating surface. Consequently, the n (dispersion coefficient) for the initial period of exposure is 0.65, and it indicates inhomogeneity of the coating surface; furthermore, it is increased with exposure periods [53].

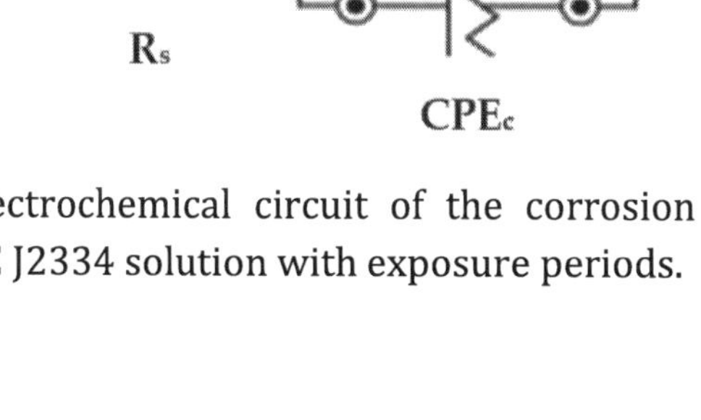

Figure 9. Electrochemical circuit of the corrosion process of Al coating in SAE J2334 solution with exposure periods.

This observation is predicting the surface becoming homogenous with the passage of exposure periods of the coating in the solution. Blucher et al. (2001) and other researchers have observed that for aluminum metal exposed in NaCl solution polluted by CO2 in the environment, the surface is inhibited by corrosion products due to the cathodic reduction of oxygen and anodic dissolution of metal by high pH [54,55,56,57]. In this study the formation of passive films that stifle the anodic dissolution of coating are as follow:

Table 5. Electrochemical parameters of Al coating extracted after the fitting of EIS data in a constant phase element (CPE) model with exposure periods in SAE J2334 solution.

Exposure Periods (h)	R_s ($\Omega \cdot cm^2$)	R_{pore} ($\Omega \cdot cm^2$)	CPE_c Y_o ($\Omega^{-1} \cdot cm^{-2} \cdot s^{-n}$)	n
1	64.15	2327.0	3.73×10^{-4}	0.65
24	62.16	2985.1	3.70×10^{-4}	0.66
144	60.65	4005.4	3.59×10^{-4}	0.67
216	62.44	4646.4	3.47×10^{-4}	0.73
384	69.55	7357.9	2.85×10^{-4}	0.80

$$2Al + NaHCO_3 + CaCl_2 + NaCl + 4H_2O + \frac{3}{2}O_2 \rightarrow NaAlCO_3(OH)_2(Dawsonite) +$$

$$Al(OH)_3(Bayerite/Gibbsite) + NaOH + CaO + 3HCl \cdot \tag{2}$$

From the above results it is very important to study the characteristics of passive film, which is being formed on the coating surface during exposure in solution. The morphology of passive film by SEM and nature of this film by XRD and Raman spectroscopy were performed. The XRD was carried out on passive film (plot not shown), which has been formed on the coating surface but was not able to detect the phases present in the passive film. It may be due to the very thin layer of passive film, which may be in nanometers, and using this technique it is not possible to detect the phases available in passive film.

3.3. Characteristics of Passive Film Formed on the Coating Surface

3.3.1. SEM Studies of Passive Film

The SEM images and EDS spectra of passive film which has been formed on coating surface in the SAE J2334 solution after 384 h of exposure are shown in Figure 10a,b. From Figure 10a it can be seen that the morphology of passive film is uniform, layered and dense [44]. The growth/orientation of passive film is prohibiting the penetration of other aggressive gases and ions. The presence of C, O, Na and Al in EDS spectra (Figure 10b) predicts the phases related to these elements present on passive film. For the evaluation of phases present in passive film, Raman spectra were carried out. For knowing the characteristics of passive film, Raman spectroscopy is a very useful technique and is described in the subsequent paragraph.

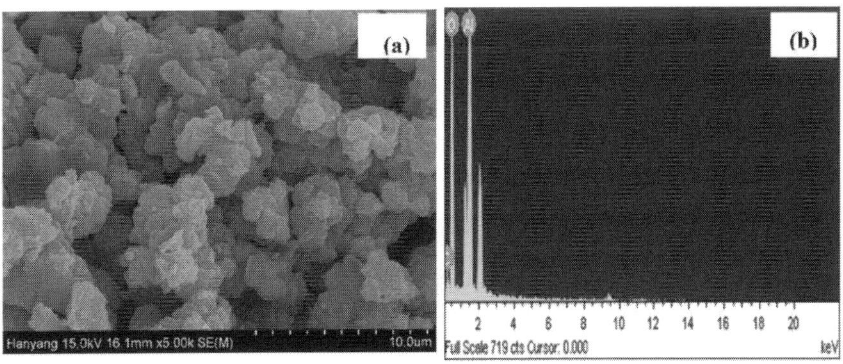

Figure 10. Morphology (a) SEM image and (b) EDS of passive film formed on the Al coating surface after 384 h of exposure in SAE J2334 solution.

3.3.2. Raman Spectroscopy of Passive Film

Raman spectroscopy of passive film was carried out after 384 h of exposure in the SAE J2334 solution and shown in Figure 11. The different phases are observed at different Raman shift, i.e., 570 cm−1 (Gibbsite) [58], 868 cm−1 (Bayerite) [58], 662 cm−1, 1405 cm−1 and 1525 cm−1 (Dawsonite) [59]. The intensity of the Raman spectra is weak due to the very low concentrations of phases and film thickness being very low. Therefore, the Raman peaks are not intense. From Raman spectroscopy it is observed that the passive film contains Bayerite (α-Al(OH)3), Gibbsite (γ-Al(OH)3) and Dawsonite (NaAlCO3(Al)2). These phases are very protective in nature and once these phases are formed on coating surface they protect the materials from further corrosion. The presence of these phases on the surface of the coating contributed enhanced corrosion resistance properties to the Al coating exposed in the SAE J2334 solution [60,61].

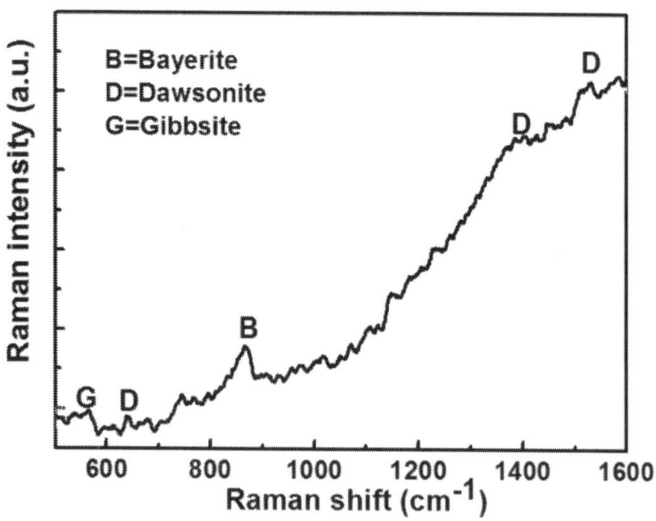

Figure 11. Raman spectroscopy of passive film formed on Al coating after 384 h of exposure in SAE J2334 solution.

Dawsonite is thermodynamically very stable and its solubility in any environment is much less [61]. The presence of Dawsonite, Gibbsite and Bayerite on the coating surface after exposure for 384 h in the SAE J2334 solution justify the findings of the above-mentioned results.

4. CONCLUSIONS

The following conclusions can be drawn from the above results:

The corrosion characteristics of the Al coating applied using the arc thermal metal spray process in the presence of stimulants provide enhanced corrosion resistance properties to the steel substrate with exposure periods in the SAE J2334 solution;

The enhanced corrosion resistance properties of the Al coating is due to presence of NaCl, CaCl2 and NaHCO3 in the solution, which reduces the anodic surface area of the coating and the cathodic reduction of oxygen by forming a thin layer of protective passive film;

Kinetics and the mechanism determined by EIS and potentiodynamic studies suggested the passivating nature of the Al coating in the presence of carbonate and chloride ions of the SAE J2334 solution with exposure periods;

The morphology of passive films is very dense, layered, thick, compact and adherent to the Al coating surface;

Raman spectroscopy confirmed the formation of Dawsonite, Gibbsite and Bayerite on the Al coating surface. These phases are very protective in nature and sparingly soluble.

ACKNOWLEDGMENTS

This research was supported by basic science research program through the National Research Foundation (NRF) of Korea funded by the Ministry of Science, ICT and Future Planning (No. 2015R1A5A1037548).

AUTHOR CONTRIBUTIONS

J.K.S. conducted the experiments and wrote the initial draft of the manuscript. H.S. Lee designed the experiments. J.K.S. and H.S. Lee analyzed the data and wrote the final manuscript. Mohamed A. Ismail and C. Bhattacharya reviewed and contributed to the final revised manuscript. All authors contributed to the analysis of the data and read the final paper.

REFERENCES

1. Jandin, G.; Liao, H.; Feng, Z.Q.; Coddet, C. Correlations between Operating Conditions, Microstructure and Mechanical Properties of Twin Wire Arc Sprayed Steel Coatings. Mater. Sci. Eng. A 2003, 349, 298–305.

2. Guilemany, J.M.; Miguel, J.M.; Armada, S.; Vizcaino, S.; Climent, F. Use of scanning white light interferometry in the characterization of wear mechanisms in thermal-sprayed coatings. Mater. Charact. 2001, 47, 307–314.

3. Rhys-Jones, T.N. Thermally sprayed coating systems for surface protection and clearance control applications in aero engines. Surf. Coat. Technol. 1990, 43–44, 402–415.

4. Pawlowski, L. The Science and Engineering of Thermal Spray Coatings, 2nd ed.; John Wiley & Sons Ltd.: West Sussex, UK, 2008.

5. Chaliampalias, D.; Vourlias, G.; Pavlidou, E.; Stergioudis, G.; Skolianos, S.; Chrissafis, K. High temperature oxidation and corrosion in marine environments of thermal spray deposited coatings. Appl. Surf. Sci. 2008, 255, 3104–3111.

6. Choe, H.-B.; Lee, H.-S.; Shin, J.-H. Experimental study on the electrochemical anti corrosion properties of steel structures applying the arc thermal metal spraying method. Materials 2014, 7, 7722–7736.

7. Krepski, R.P. Thermal Spray: Coating Applications in the Chemical Process Industries; Published for the Materials Technology Institute of the Chemical Process Industries Inc.: St. Louis, MO, USA, 1993; Volume 2, p. 8.

8. Paredes, R.S.C.; Amico, S.C.; d'Oliveira, A.S.C.M. The effect of roughness and pre-heating of the substrate on the morphology of aluminium coatings deposited by thermal spraying. Surf. Coat. Technol. 2006, 200, 3049–3055.

9. Li, Y.; Liu, J.; Duan, J.; Hou, B. Thermally sprayed aluminium and zinc coatings for tidal zone cathodic protection of offshore platform pile leg. Mater. Perform. 2006, 45, 16–20.

10. Muhamad, H.A.M.; Hayati, S.N.; Kiyai, A.S.; Binti, M.S.N. Thermal arc spray overview. Mater. Sci. Eng. 2013, 46, 1–10.

11. Jiang, Q.; Miao, Q.; Liang, W.-P.; Ying, F.; Tong, F.; Xu, Y.; Ren, B.-L.; Yao, Z.-J.; Zhang, P.-Z. Corrosion behavior of arc sprayed Al–Zn–Si–RE coatings on mild steel in 3.5 wt. % NaCl solution. Electrochim. Acta 2014, 115, 644–656.

12. Valdez, B.; Kiyota, S.; Stoytcheva, M.; Zlatev, R.; Bastidas, J.M. Cerium-based conversion coatings to improve the corrosion

resistance of aluminium alloy 6061-T6. Corros. Sci. 2014, 87, 141–149.

13. Panossian, Z.; Mariaca, L.; Morcillo, M.; Flores, S.; Rocha, J.; Pena, J.J.; Herrera, F.; Corvo, F.; Sanchez, M.; Rincon, O.T.; et al. Steel Cathodic Protection Afforded by Zinc, Aluminium and Zinc/Aluminium Alloy Coatings in the Atmosphere. Surf. Coat. Technol. 2005, 190, 244–248.

14. Schweitzer, P.A. Atmospheric Corrosion. In Fundamentals of Corrosion Mechanisms, Causes and Preventive Methods; CRC Press, Taylor and Francis Group: Boca Raton, FL, USA, 2009; Chapter 4.

15. Salas, O.; de Troconis, O.; Rojas, D.; Tosaya, A.; Nathalie, R.; Miguel, S. Six-year evaluation of thermal-sprayed coating Zn/Al in tropical marine environments. Int. J. Corros. 2012.

16. Steffens, H.D.; Babiak, Z.; Wewel, M. Recent developments in arc spraying. IEEE Trans. Plasma Sci. 1990, 18, 974–975.

17. Davis, J.R. Surface Engineering for Corrosion and Wear Resistance; ASM International: Materials Park, OH, USA, 2001.

18. Muhamad, H.A.M.; Hayati, S.N.; Kiyai, A.S.; Binti, M.S.N. Critical process and performance parameters of thermal arc spray coating. Int. J. Mater. Eng. Innov. 2014, 5, 12–27.

19. ASTM D4541-09e1, Standard Test Method for Pull-Off Strength of Coatings Using Portable Adhesion Testers; ASTM International: West Conshohocken, PA, USA, 2009.

20. Standard SAE J2334; Society of Automotive Engineers (SAE): Warrendale, PA, USA, 2003.

21. Saha, J.K.; Mitra, P.K.; Paul, S.; Singh, D.D.N. Performance of different organic coatings on steel substrate by accelerated and

in atmospheric exposure tests. Indian J. Chem. Technol. 2010, 17, 102–110.

22. Singh, D.D.N.; Dey, M.; Singh, V. Role of Buffering and Complexing Agents in Zinc Plating Chloride Baths on Corrosion-Resistance of Produced Coatings. Corrosion 2002, 58, 971–980.

23. Singh, D.D.N.; Yadav, S.; Saha, J.K. Role of climatic conditions on corrosion characteristics of structural steels. Corros. Sci. 2008, 50, 93–110.

24. Townsend, H.E.; Simpson, M.W.; van der Linde, W.B.; McCune, D.C. License Plate Cosmetic Corrosion Test of Automotive Coated Steel Sheet. Corrosion 1999, 55, 406–411.

25. Abedi, E.E.; Hamidreza, S.; Mohamad, A.G.; Javad, M.; Larry, P. Study of Corrosion Behavior of Arc Sprayed Aluminum Coating on Mild Steel. J. Therm. Spray Technol. 2012, 21, 1195–1202.

26. Han, M.-S.; Woo, Y.-B.; Ko, S.-C.; Jeong, Y.-J.; Jang, S.-K.; Kim, S.-J. Effects of thickness of Al thermal spray coating for STS 304. Trans. Nonferr. Met. Soc. China 2009, 19, 925–929.

27. Torres, B.; Campo, M.; Urena, A.; Rams, J. Thermal spray coatings of highly reinforced aluminium matrix composites with sol-gel silica coated SiC particles. Surf. Coat. Technol. 2007, 201, 7552–7559.

28. Deshpande, S.; Kulkami, A.; Sampath, S.; Herman, H. Application of image analysis for characterization of porosity in thermal spray coatings and correlation with small angle neutron scattering. Surf. Coat. Technol. 2004, 187, 6–16.

29. Celik, E.; Ozdemir, I.; Avcic, E.; Tsunekawa, Y. Corrosion behaviour of plasma sprayed coatings. Surf. Coat. Technol. 2005, 193, 297–302.

30. Kawakita, J.; Kuroda, S.; Fukushima, T.; Kodama, T. Corrosion resistance of HVOF sprayed HastelloyC nickel base alloy in seawater. Corros. Sci. 2003, 45, 2819–2835.

31. Baxter, C.S. Thermal spray inspection method and training requirement. Mater. Perform. 1997, 36, 31–32.

32. Wixson, D. Thermal spray deposits shield structure from corrosion. Weld. J. 2009, 88, 46–48.

33. Salman, S.A.; Ichino, R.; Okido, M.A. Comparative electrochemical study of AZ31 and AZ91 magnesium alloy. Int. J. Corros. 2010, 2010, 412129.

34. Dean, S.W. Electrochemical methods of corrosion testing. In Electrochemical Techniques for Corrosion; Baboian, R., Ed.; NACE: Houston, TX, USA, 1977; pp. 52–60.

35. Dzhurinskiy, D.; Maeva, E.; Leshchinsky, E.; Maev, R.G. Corrosion protection of light alloys using low pressure cold spray. J. Therm. Spray Technol. 2012, 21, 304–313.

36. Moran, A.L.; Shaw, B.A. In Situ Evaluation of Oxide Formation in Aluminum Thermal Spray Coatings. J. Electrochem. Soc. 1988, 135, 2773–2774.

37. Hurlen, T.; Haijg, A.T. Corrosion and passive behavior of Aluminum in weakly alkaline solution. Electrochem. Acta 1984, 29, 1833–1838.

38. Liu, Y.; Visser, P.; Zhou, X.; Lyon, S.B.; Hashimoto, T.; Curioni, M.; Gholinia, A.; Thompson, G.E.; Smyth, G.; Gibbon, S.R.; et al. Protective Film Formation on AA2024-T3 Aluminum Alloy by Leaching of Lithium Carbonate from an Organic Coating. J. Electrochem. Soc. 2016, 163, C45–C53.

39. Ishizaki, T.; Masuda, Y.; Teshima, K. Composite film formed on magnesium alloy AZ31 by chemical conversion from

molybdate/phosphate/fluorinate aqueous solution toward corrosion protection. Surf. Coat. Technol. 2013, 217, 76–83.

40. Wen, J.B.; Ma, J.J.; He, J.G. Al-Base Sacrificial Anode Material for Corrosion Protection; Chemical Industry Press: Beijing, China, 2012; pp. 67–79.

41. Ryl, J.; Wysocka, J.; Jarzynka, M.; Zielinski, A.; Orlikowski, J.; Darowicki, K. Effect of native air-formed oxidation on the corrosion behaviour of AA 7075 aluminium alloys. Corros. Sci. 2014, 87, 150–155.

42. Onofre-Bustamante, E.; Dominguez-Crespo, M.A.; Torres-Huerta, A.M.; Olvera-Martinez, A.; Genesca-Llongueras, J.; Rodriguez-Gomez, F.J. Characterization of cerium-based conversion coatings for corrosion protection of AISI-1010 commercial carbon steel. J. Solid State Electrochem. 2009, 13, 1785–1799.

43. Stefanov, P.; Atanasova, G.; Stoychev, D.; Marinova, T.S. Electrochemical deposition of CeO2 on ZrO2 and Al2O3 thin films formed on stainless steel. Surf. Coat. Technol. 2004, 180–181, 446–449.

44. Dongrui, Y.; Omar, R.; Homero, C. FeCO3 layer evolution for API 5L X52 steel in carbon dioxide-saturated NaCl brine in the presence of 1-decyl-3-methylimidazolium chloride. Corros. Sci. 2014, 87, 40–50.

45. Verdian, M.M.; Raeissi, K.; Salehi, M. Electrochemical Impedance Spectroscopy of HVOF-Sprayed NiTi Intermetallic Coatings Deposited on AISI 1045 Steel. J. Alloy. Compd. 2010, 507, 42–46.

46. Gudic, S.; Radosevic, J.; Kliskic, M. Study of passivation of Al and Al-Sn alloys in borate buffer solutions using electrochemical impedance spectroscopy. Electrochim. Acta 2002, 47, 3009–3016.

47. Martin, F.J.; Cheek, G.T.; O'Grady, W.E.; Natishan, P.M. Impedance studies of the passive film on aluminium. Corros. Sci. 2005, 47, 3187–3201.

48. Liu, Y.J.; Wang, Z.Y.; Ke, W. Study on influence of native oxide and corrosion products on atmospheric corrosion of pure Al. Corros. Sci. 2014, 80, 169–176.

49. Maddela, S.; O'Keefe, M.J.; Wang, Y.M.; Kuo, H.H. Influence of surface pretreatment on coating morphology and corrosion performance of cerium-based conversion coatings on AZ91D alloy. Corrosion 2010, 66, 1–8.

50. Bastidas, J.M.; Polo, J.L.; Torres, C.L.; Cano, E. A study on the stability of AISI 316L stainless steel pitting corrosion through its transfer functions. Corros. Sci. 2001, 43, 269–281.

51. Flores, J.F.; Olaya, J.J.; Colas, R.; Rodil, S.E.; Valdez, B.S.; Fuente, I.G. Corrosion behaviour of TaN thin PVD films on steels. Corros. Eng. Sci. Technol. 2006, 41, 168–176.

52. Cox, B.; Wong, Y.M. Simulation porous oxide films on Zirconium alloys. J. Nucl. Mater. 1995, 218, 324–334.

53. Singh, J.K.; Singh, D.D.N. The nature of rusts and corrosion characteristics of low alloy and plain carbon steels in three kinds of concrete pore solution with salinity and different pH. Corros. Sci. 2012, 56, 129–142.

54. Blucher, D.B.; Lindstrom, R.; Svensson, J.E.; Johansson, L.G. The Effect of CO2 on the NaCl-Induced Atmospheric Corrosion of Aluminum. J. Electrochem. Soc. 2001, 148, B127–B131.

55. Lin, C.; Li, X. Role of CO2 in the initial stage of atmospheric corrosion of AZ91 magnesium alloy in the presence of NaCl. Rare Met. 2006, 25, 190–196.

56. Maier, B.; Frankel, G.S. Behavior of Magnesium-Rich Primers on AA2024-T3. Corrosion 2011, 67, 1–15.

57. Pathak, S.S.; Blanton, M.D.; Mendon, S.K.; Rawlins, J.W. Mineralogical Transformation and Electrochemical Nature of Magnesium-Rich Primers during Natural Weathering. Metals 2014, 4, 322–334.

58. Ruan, H.D.; Frost, R.L.; Kloprogg, J.T. Comparison of the Raman spectra of Bayerite, Boehmite, Diaspore and Gibbsite. J. Raman Spectrosc. 2001, 32, 745–750.

59. Frost, R.L.; Bouzaid, J.M. Raman spectroscopy of Dawsonite NaAl(CO3)(OH)2. J. Raman Spectrosc. 2007, 38, 873–879.

60. Pascale, B.; Palmer, D.A.; Anovitz, L.M.; Juske, H. Dawsonite synthesis and re-evaluation of its thermodynamic properties from solubility measurements: Implications for mineral trapping of CO2. Geochim. Cosmochim. Acta 2007, 71, 4438–4455.

61. Ferrante, M.J.; Stuve, J.M.; Richardson, D.W. Thermodynamic Data for Synthetic Dawsonite; Bureau of Mines: Washington, DC, USA, 1976; p. 13.

CHAPTER 2

Fabrication of Corrosion Resistance Micro-Nanostructured Superhydrophobic Anodized Aluminum in a One-Step Electrodeposition Process

Ying Huang, Dilip K. Sarkar * and X.-Grant Chen

Centre Universitaire de Recherche sur l'Aluminium (CURAL), Université du Québec à Chicoutimi, 555 Boulevard de l'Université, Chicoutimi, QC G7H 2B1, Canada

ABSTRACT

The formation of low surface energy hybrid organic-inorganic micro-nanostructured zinc stearate electrodeposit transformed the anodic aluminum oxide (AAO) surface to superhydrophobic, having a water contact angle of 160°. The corrosion current densities of the anodized and aluminum alloy surfaces are found to be 200 and 400 nA/cm^2, respectively. In comparison, superhydrophobic anodic aluminum oxide (SHAAO) shows a much lower value of 88 nA/cm^2. Similarly, the charge transfer resistance, R_{ct}, measured by electrochemical impedance spectroscopy shows that the SHAAO substrate was found to be 200-times larger than the as-received aluminum alloy substrate. These results proved that the superhydrophobic surfaces created on the anodized surface significantly improved the corrosion resistance property of the aluminum alloy.

KEYWORDS

superhydrophobic aluminum; corrosion; anodized aluminum oxide (AAO); organic-inorganic; micro-nanostructure; zinc stearate (ZnSA); potentiodynamic polarization; electrochemical impedance spectroscopy (EIS)

1. INTRODUCTION

Aluminum (Al) and its alloys are naturally-abundant engineering materials with extensive applications in daily life. In recent years, nanoporous anodic aluminum oxide (AAO) prepared by electrochemical anodization has found a multitude of applications, such as catalysis [1], drug delivery [2], biosensing [3], template synthesis [4], molecular and ion separation [5], corrosion resistance [6], and so forth. The formation of AAO on the aluminum alloy surface would act as the corrosion barrier. The formation of AAO was limited to a certain extent due to its hydrophilic behavior. Therefore, it is necessary to transform the AAO to be superhydrophobic in order to improve the corrosion resistance properties.

Superhydrophobicity, exhibiting an excellent water-repellent property, is characterized by a contact angle above 150°. Creating a rough surface, as well as reducing the surface energy is attributed to the modification of superhydrophobicity [7]. In the last few decades, a large effort has been devoted to the realization of superhydrophobic surfaces, due to their applications in biology [8], anti-corrosion [9,10,11,12], anti-icing [13], self-cleaning [14], *etc.*

Recent publications show that the anodized surface can be made superhydrophobic by passivation with organic molecules [15,16,17,18]. In the study of Liu *et al.*, the superhydrophobic anodized surfaces were fabricated by polypropylene (PP) coating after anodizing [17]. Li *et al.*, used a very complex process to engineer superhydrophobic anodized aluminum alloy surfaces [16]. After anodizing with sulfuric acid, the anodized sample was firstly immersed in the mixing solution containing $M(NO_3)_2$ salt (M = Mg, Co, Ni and Zn) and NH_4NO_3, followed by immersion in (Heptadecafluoro-1,1,2,2-tetradecyl)trimethoxysilane (*n*-$CF_3(CF_2)_7CH_2CH_2Si(OC_2H_5)_3$). Vengatesh and Kulandainathan fabricated superhydrophobic anodic aluminum oxide (SHAAO) surfaces by passivation with organic molecules and show these surfaces having corrosion resistance properties indicated by potentiodynamic polarization [15]. Despite this, the fabrication, as well as corrosion resistance properties of the electrodeposited superhydrophobic surfaces on AAO have not been shown in the literature.

In this work, we have described the method to prepare the superhydrophobic surfaces on AAO by the electrodeposition process and describe their corrosion resistance properties, both using potentiodynamic polarization and electrochemical impedance spectroscopy (EIS).

2. EXPERIMENTAL PROCEDURE

After pretreatment with 0.01 M NaOH at 55 °C for 3 min, the AA6061 aluminum alloy substrates were anodized using 3 vol. % H_3PO_4 aqueous solution at 10 °C at 0.01 A/cm^2 (Ametek Sorensen DCS 100-12E, Chicoutimi, QC, Canada) for 2 h. The superhydrophobic anodic aluminum oxide (SHAAO) surface was prepared by the electrodeposition process in an ethanolic solution containing 0.01 M stearic acid (SA) and 0.01 M zinc nitrate ($Zn(NO_3)_2$) by applying 20 V (Ametek Sorensen DCS 100-12E) for 10 min. Microstructural examination was conducted using a scanning electron microscope (SEM, JEOL JSM-6480 LV, Chicoutimi, QC, Canada). Surface roughness was measured using an optical profilometer. The X-ray diffraction (XRD) analyses of the samples were carried out using a Bruker D8 Discover system (Chicoutimi, QC, Canada). The chemical composition of the samples was analyzed by means of Fourier transform infrared spectroscopy (FTIR, Perkins Elmer Spectrum One, Chicoutimi, QC, Canada) and an energy dispersive X-ray spectrometer (EDX, JEOL JSM-6480 LV, Chicoutimi, QC, Canada). Wetting characteristics of sample surfaces were evaluated by measuring static contact angles (CA) using a first ten Angstrom contact angle goniometer at five positions on each substrate using a 10-μL deionized water drop. In the case of rolling-off surfaces, the contact angle was measured by holding the water drop between the needle and the surface, as presented in the inset image of Figure 1c. Electrochemical experiments were performed using a PGZ100 potentiostat and a 300-cm^3-EG & G PAR flat cell (London Scientific, London, ON, Canada), equipped with a standard three-electrode system with an Ag/AgCl reference electrode, a platinum (Pt) mesh as the counter electrode (CE) and the sample as the working electrode (WE). Before the test, the open circuit potential (OCP) was monitored for more than 20 h for stabilization by immersing the sample surface in 3 wt. % NaCl aqueous solution. The electrochemical impedance spectroscopy (EIS) was tested in the frequency range between 10

MHz and 100 kHz with a sine-wave amplitude of 10 mV. For the potentiodynamic polarization experiments, the potential was scanned from −250 mV to +1000 mV with respect to the OCP voltage at a scan rate of 2 mV/s.

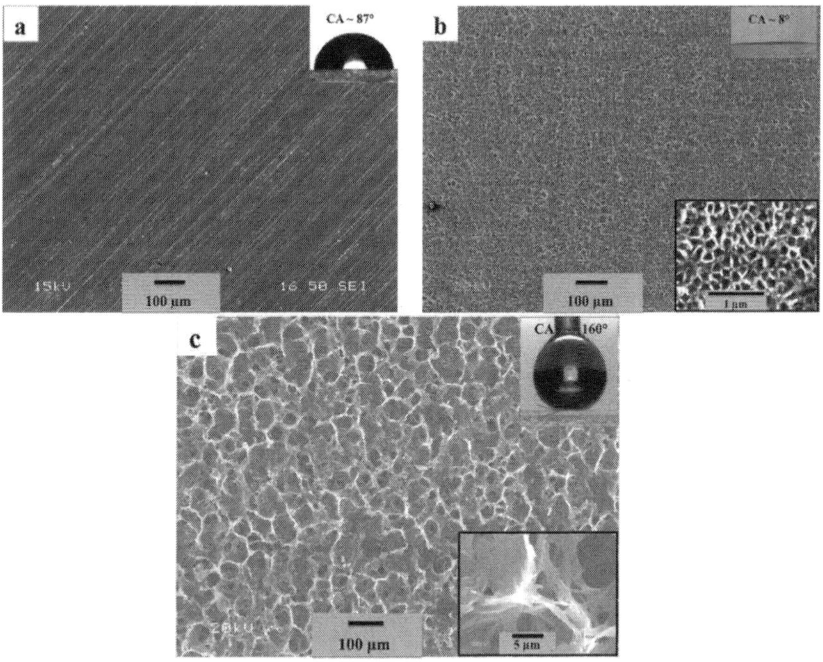

Figure 1. Secondary electron SEM micrographs showing the top surface of (**a**) the as-received aluminum alloy; (**b**) the anodized and (**c**) the electrodeposited anodized substrates. The inset top-right images present the water drop on the surfaces, and the bottom-right ones show magnified microstructures.

3. RESULTS AND DISCUSSION

SEM images in Figure 1 reveal the evolution of the different morphologies of: (a) the as-received the aluminum alloy surface; (b) the anodized (AAO) surface; and (c) the electrodeposited hybrid organic-inorganic anodized surface using the ethanolic electrolytic solution containing stearic acid (SA) and zinc nitrate ($Zn(NO_3)_2$) at 20 V DC. The as-received aluminum alloy surface was characterized by parallel lines resulting from the rolling process, corresponding to a surface roughness of 0.45 ± 0.03 μm and a contact angle of 87° ± 3° (Figure 1a).

After anodizing in phosphoric acid for a duration of 2 h, uniformly-distributed nanopores with an average diameter of approximately 100 nm and an inter-pore distance of ~137 nm are observed on the surface of the anodized substrate (Figure 1b), with a roughness of 0.68 ± 0.02 μm, while the contact angle was measured to be 8° ± 1°, indicating a superhydrophilic property. It is however evident from Figure 1c that the electrodeposition process resulted in the appearance of a porous network microstructure on the anodized substrate. This substrate was built by nanofibre clusters connected with each other, as presented in the inset of Figure 1c. It can be also observed that these micro-nanoporous structures are distributed uniformly on the anodized surface, resulting in a micro-nanorough surface having a roughness of 6.85 ± 1.02 μm. Interestingly, this surface shows superhydrophobic properties with a contact angle of 160° ± 1° having a contact angle hysteresis of 2° ± 1°.

Low angle X-ray diffraction (XRD), energy dispersive X-ray spectroscopy (EDX) and Fourier transform infrared spectroscopy (FTIR) have been carried out to determine the composition of the electrodeposited micro-nanostructure thin films on the anodized substrate, as shown in Figure 2. Figure 2a(a3) shows four distinct peaks at 4.2°, 6.26°, 8.3° and 10.4°, which correspond to zinc stearate

($(CH_3(CH_2)_{16}COO)_2Zn$) (abbreviated as ZnSA,). The possible mechanism of the formation of ZnSA has been presented as follows:

$$2CH_3(CH_2)_{16}COOH + Zn^{2+} \xrightarrow{20\,V} (CH_3(CH_2)_{16}COO)_2Zn + 2H^+$$

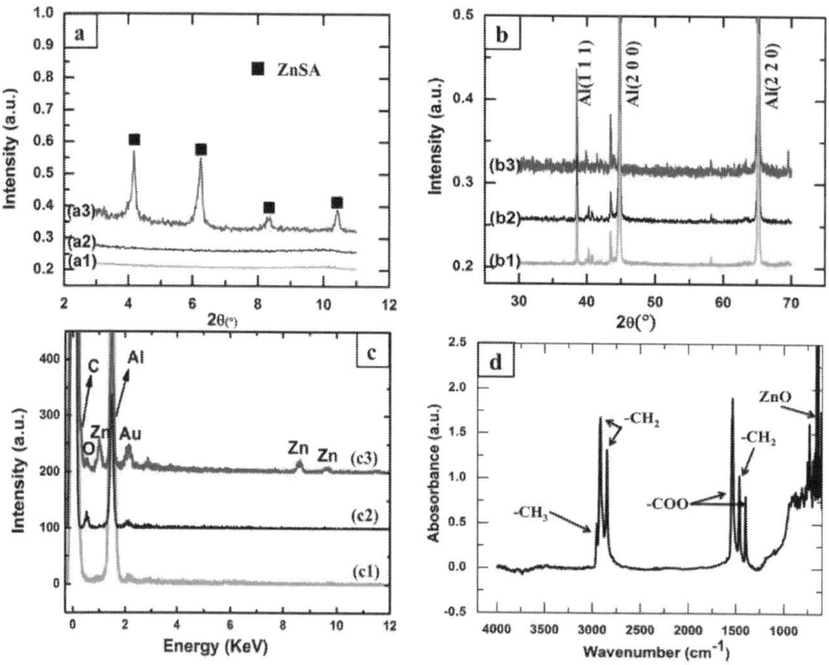

Figure 2. (**a**) Low angle XRD patterns of (**a1**) as-received aluminum alloy, (**a2**) AAO, (**a3**) electrodeposited anodized substrate; (**b**) high angle XRD patterns of (**b1**) as-received aluminum alloy, (**b2**) AAO and (**b3**) electrodeposited anodized substrate; (**c**) EDX spectra of (**c1**) the as-received aluminum alloy, (**c2**) anodized and (**c3**) electrodeposited anodized substrate (ZnSA = zinc stearate); (**d**) FTIR spectrum of the electrodeposited anodized substrate.

This reaction mechanism is very similar to that mentioned by Liu *et al.* for the electrodeposition of cerium myristate by electrodeposition on magnesium substrates to obtain superhydrophobicity [19]. When the DC voltage was applied on the electrodes, the Zn^{2+} ions close to the cathodic electrode reacted with SA, forming ZnSA and H^+ ions, as presented in Equation (1). Meanwhile, some of the H^+ ions obtained an electron and formed H_2 gas. It is noted that the as-received aluminum alloy and anodized substrates of Figure 2a(a1,a2) do not show any characteristic peaks. Furthermore, in Figure 2b, it shows the peaks at 38.47°, 44.72° and 65.1°, respectively, which are in good agreement with the characteristic peaks of Al(111), (200) and (220). This arises from the substrate of the aluminum alloy.

EDX analysis (Figure 2c) revealed that the chemical composition of the anodized surface consisted of O and Al, resulting from the formation of aluminum oxide (Al_2O_3), whereas only the Al peak was seen in the spectrum of the as-received aluminum alloy surface. However, C, Zn, O and Al are observed in the spectrum of electrodeposited anodized surface, confirming the formation of ZnSA complementary with the XRD pattern of the electrodeposited anodized substrate. It is worth mentioning that the Au peaks appearing in Figure 2c are due to the thin layer of gold on the electrodeposited thin films for improving the resolution by eliminating the charging effect of non-conducting samples during EDX analyses.

In the FTIR spectrum of the electrodeposited hybrid organic-inorganic anodized substrate (Figure 2d), the appearance of the –CH group (–CH_3 at 2962 cm^{-1}, as well as –CH_2 at 1459 cm^{-1}, 2850 cm^{-1} and 2919 cm^{-1}), as well as –COO at 1395 and 1550 cm^{-1} indicated the formation of ZnSA on the surface [20], which is in good agreement with the XRD and EDX results. These results support the formation of low surface energy methylated (–CH_3 and –CH_2) components on the electrodeposited anodized surface and make it superhydrophobic. In addition, the ZnO peaks at 560 cm^{-1} may come from the bonding of –COOZn, which further verifies the reaction production of ZnSA formed on the modified surface in a one-step electrodeposition process.

Figure 3 illustrates the variation of open circuit potential (OCP) on the superhydrophobic AAO (SHAAO) surface. The OCP value shifted from −177 to −708 mV with an average of −281 ± 87 mV into a 5-h immersion time in the salt solution. It is quite unstable during this period. From 5 to 16 h of immersion time, the OCP value varied from −533 to −686 mV with an average of −610 ± 26.8 mV. With the prolongation of immersion time to 20 h, the OCP value varied from −708 to −730 mV, with an average of −717 ± 2.8 mV. It can be seen that the fluctuation of OCP voltage reduces with time and stabilized nearly after 20 h. In contrast, the surface of the aluminum alloy substrate gets stabilized within 30 min of immersion time in the salt solution. The OCP fluctuations of the SHAAO substrate may be due to the poor wetting, as well as protective properties of the SHAAO surface with the salt solution. Therefore, the EIS and polarization experiments were performed after 20 h of immersion of the superhydrophobic surface in the salt solution while monitoring the OCP continuously.

To evaluate the corrosion resistance performance of the fabricated anodized and SHAAO substrates, potentiodynamic polarization and electrochemical impedance spectroscopy (EIS) tests were carried out. Figure 4a shows the polarization curves of the as-received aluminum alloy, anodized and SHAAO substrates, respectively. The as-received aluminum alloy substrate exhibited a corrosion current density (I_{corr}) of 400 nA/cm^2 and a polarization resistance (R_p) of 50 kΩ·cm^2. The R_p value was calculated by the Stern-Geary equation, as shown in Equation (2).

$$Rp = \frac{\beta_a \beta_c}{2.3 I_{corr}(\beta_a + \beta_c)} \tag{2}$$

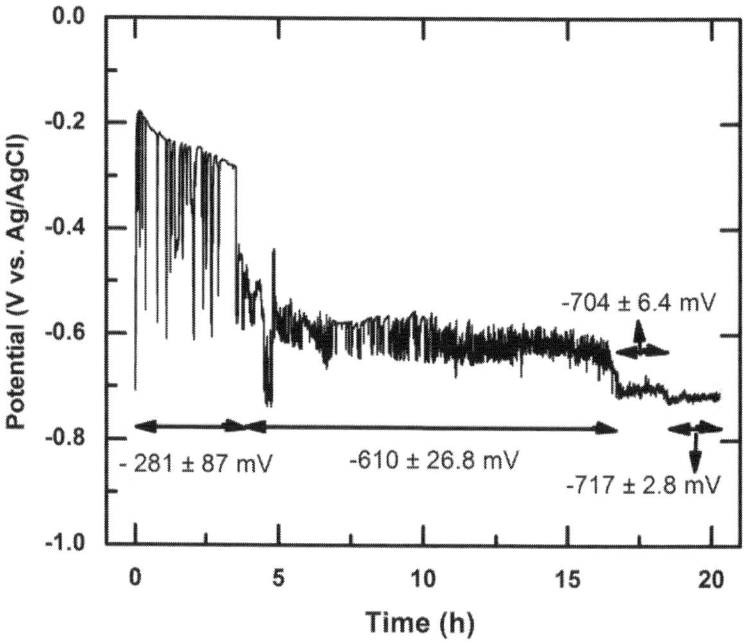

Figure 3. The variation of open circuit potential (OCP) with time on the superhydrophobic AAO (SHAAO) surface.

When the aluminum alloy substrate was anodized, the I_{corr} was reduced to 200 nA/cm^2, and R_p increased to 87 kΩ·cm^2, indicating an improved corrosion resistance compared to the as-received substrate. This was due to a barrier layer formed on the anodized surface. On the other hand, the I_{corr} of the hybrid organic-inorganic SHAAO substrate decreased remarkably to 88 nA/cm^2, while the R_p increased up to 441 kΩ·cm^2. This shows an even better anti-corrosion performance of the SHAAO substrate relative to the anodized substrate, likely attributed to the superhydrophobic ZnSA coating formed on the surface. It is noticed that the current density of the SHAAO substrate was increased sharply at around −0.4 V *vs.* Ag/AgCl, which might have been related to the dissolution of the superhydrophobic film with the prolongation of corrosion time (more than 20 h). It is also found that the current density of the SHAAO substrate at −0.4 V to 0.2 V *vs.* Ag/AgCl was parallel to, but smaller than, that of the anodized substrate. This might be due to the superhydrophobic material ZnSA filling the anodized pore structure. As a result, the SHAAO substrate after dissolution of the surface material still presents a good corrosion resistance property.

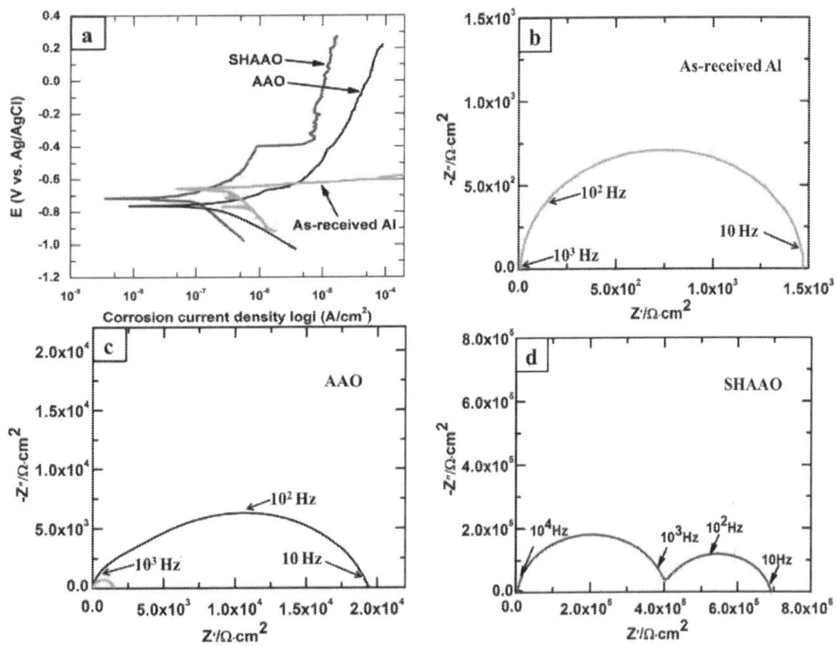

Figure 4. (**a**) Polarization curves of the as-received aluminum alloy, AAO and SHAAO substrates in 3 wt. % NaCl corrosive solution; (**b–d**) the Nyquist plots of (**b**) as-received aluminum alloy; (**c**) AAO (as-received aluminum alloy (a small semicircle in green close to the origin) also shown for comparison) and (**d**) SHAAO substrate (as-received aluminum alloy and AAO also shown inside for comparison, very small semicircle, nearly visible, close to the origin).

The I_{corr} of the electrodeposited cerium stearate superhydrophobic Mg surfaces was reported to be 142 nA/cm^2 with a 30-min immersion to perform OCP [19]. Experiments performed by Vengatesh and Kulandainathan with a 30-min immersion before polarization show that the corrosion current varies between 2 and 1050 nA/cm^2, depending on the passivated molecules [15]. Moreover, in the study of He *et al.*, a 1-h immersion time is used for the stabilization [21]. In our experiment, the samples were exposed to NaCl solution for 20 h to stabilize under OCP. This indicates that our superhydrophobic film displays a better stability and durability in the corrosion test as compared to the reported values in the literature.

Nyquist plots of the as-received aluminum alloy, anodized and hybrid organic-inorganic SHAAO substrate are presented in Figure 4b–d. It is well known that R_{ct} is the charge transfer resistance, describing the difficulty of the corrosion occurring on the substrate. R_{ct} of anodized substrate is found to be 14 kΩ·cm^2 in Figure 4c, which is about 10-times higher than that of the as-received aluminum alloy substrate (1.5 kΩ·cm^2) in Figure 4b. However, the corrosion protection of the SHAAO substrate is the combined effect of R_{ct} of 284 kΩ·cm^2 (semicircle at low frequency) along with the resistance of the superhydrophobic films of 405 kΩ·cm^2 (semicircle at high frequency) in Figure 4d. This indicates a significant enhancement of corrosion resistance, which is complementary to the result from the polarization curves, where the polarization resistance of SHAAO is higher than the anodized substrate, as well as for the Al substrate.

In the study of Liu *et al.* [19], the R_{ct} of the superhydrophobic magnesium substrate covered with cerium myristate was found to be 13 kΩ, which is comparable to our anodized substrate, but much smaller than our zinc stearate-covered SHAAO substrate [19].

4. CONCLUSIONS

A superhydrophobic anodic aluminum oxide (SHAAO) surface was prepared by the electrodeposition process using the ethanolic solution of stearic acid (SA) mixed with zinc nitrate ($Zn(NO_3)_2$) at a constant voltage of 20 V. The hybrid organic-inorganic SHAAO surface having a micro-nanoporous structure of low surface energy zinc stearate (ZnSA) exhibits a water contact angle (CA) of 160° ± 1°. The SHAAO substrate has a polarization resistance (R_p) and charge transfer resistance (R_{ct}) of 441 and 284 kΩ·cm², respectively, much higher than that of the as-received aluminum(R_p of 50 kΩ·cm² and R_{ct} of 1.46 kΩ·cm²) and anodized aluminum substrate (R_p of 87 kΩ·cm² and R_{ct} of 14 kΩ·cm²) in the corrosion. This indicates that the SHAAO substrate displays a much better corrosion resistance property as compared to the as-received aluminum alloy substrate, as well as the anodic aluminum oxide (AAO) substrate.

ACKNOWLEDGMENTS

We acknowledge the financial support provided by the Natural Science and Engineering Research Council of Canada (NSERC) and the Aluminium Research Centre—REGAL.

AUTHOR CONTRIBUTIONS

Ying Huang performed experiments, analysed data and wrote the draft of manuscript; D.K. Sarkar and X.-Grant Chen supervised the works. All authors took part in the discussion, the preparation and revision of the manuscript.

REFERENCES

1. Dotzauer, D.M.; Dai, J.; Sun, L.; Bruening, M.L. Catalytic membranes prepared using layer-by-layer adsorption of polyelectrolyte/metal nanoparticle films in porous supports. *Nano Lett.* **2006**, *6*, 2268–2272.

2. Simovic, S.; Losic, D.; Vasilev, K. Controlled drug release from porous materials by plasma polymer deposition. *Chem. Commun.* **2010**, *46*, 1317–1319.

3. Wang, M.; Meng, G.; Huang, Q.; Xu, Q.; Chu, Z.; Zhu, C. FITC-modified PPy nanotubes embedded in nanoporous AAO membrane can detect trace PCB20 via fluorescence ratiometric measurement. *Chem. Commun.* **2011**, *47*, 3808–3810.

4. Liu, Y.; Goebl, J.; Yin, Y. Templated synthesis of nanostructured materials. *Chem. Soc. Rev.* **2013**, *42*, 2610–2653.

5. Jani, A.M.M.; Anglin, E.J.; McInnes, S.J.P.; Losic, D.; Shapter, J.G.; Voelcker, N.H. Nanoporous anodic aluminium oxide membranes with layered surface chemistry. *Chem. Commun.* **2009**, *21*, 3062–3064.

6. Bouchama, L.; Azzouz, N.; Boukmouche, N.; Chopart, J.P.; Daltin, A.L.; Bouznit, Y. Enhancing aluminum corrosion resistance by two-step anodizing process. *Surf. Coat. Technol.* **2013**, *235*, 676–684.

7. Huang, Y.; Sarkar, D.K.; Chen, X.-G. A one-step process to engineer superhydrophobic copper surfaces. *Mater. Lett.* **2010**, *64*, 2722–2724.

8. Ciasca, G.; Papi, M.; Chiarpotto, M.; de Ninno, A.; Giovine, E.; Campi, G.; Gerardino, A.; de Spirito, M.; Businaro, L. Controlling the cassie-to-wenzel transition: An easy route towards the

realization of tridimensional arrays of biological objects. *Nano-Micro Lett.* **2014**, *6*, 280–286.

9. Zhang, F.; Chen, S.; Dong, L.; Lei, Y.; Liu, T.; Yin, Y. Preparation of superhydrophobic films on titanium as effective corrosion barriers. *Appl. Surf. Sci.* **2011**, *257*, 2587–2591.

10. Yin, B.; Fang, L.; Tang, A.-Q.; Huang, Q.-L.; Hu, J.; Mao, J.-H.; Bai, G.; Bai, H. Novel strategy in increasing stability and corrosion resistance for super-hydrophobic coating on aluminum alloy surfaces. *Appl. Surf. Sci.* **2011**, *258*, 580–585.

11. Huang, Y.; Sarkar, D.K.; Chen, X.-G. Superhydrophobic aluminum alloy surfaces prepared by chemical etching process and their corrosion resistance properties. *Appl. Surf. Sci.* **2015**, *356*, 1012–1024.

12. Huang, Y.; Sarkar, D.K.; Gallant, D.; Chen, X.-G. Corrosion resistance properties of superhydrophobic copper surfaces fabricated by one-step electrochemical modification process. *Appl. Surf. Sci.* **2013**, *282*, 689–694.

13. Zuo, Z.; Liao, R.; Guo, C.; Yuan, Y.; Zhao, X.; Zhuang, A.; Zhang, Y. Fabrication and anti-icing property of coral-like superhydrophobic aluminum surface. *Appl. Surf. Sci.* **2015**, *331*, 132–139.

14. Watson, G.S.; Green, D.W.; Schwarzkopf, L.; Li, X.; Cribb, B.W.; Myhra, S.; Watson, J.A. A gecko skin micro/nano structure—A low adhesion, superhydrophobic, anti-wetting, self-cleaning, biocompatible, antibacterial surface. *Acta Biomater.* **2015**, *21*, 109–122.

15. Vengatesh, P.; Kulandainathan, M.A. Hierarchically ordered self-lubricating superhydrophobic anodized aluminum surfaces with enhanced corrosion resistance. *ACS Appl. Mater. Interfaces* **2015**, *7*, 1516–1526.

16. Li, Y.; Li, S.; Zhang, Y.; Yu, M.; Liu, J. Fabrication of superhydrophobic layered double hydroxides films with different metal cations on anodized aluminum 2198 alloy. *Mater. Lett.* **2015**, *142*, 137–140.

17. Liu, W.; Luo, Y.; Sun, L.; Wu, R.; Jiang, H.; Liu, Y. Fabrication of the superhydrophobic surface on aluminum alloy by anodizing and polymeric coating. *Appl. Surf. Sci.* **2013**, *264*, 872–878.

18. Fujii, T.; Aoki, Y.; Habazaki, H. Superhydrophobic hierarchical surfaces fabricated by anodizing of oblique angle deposited Al–Nb alloy columnar films. *Appl. Surf. Sci.* **2011**, *257*, 8282–8288.

19. Liu, Q.; Kang, Z. One-step electrodeposition process to fabricate superhydrophobic surface with improved anticorrosion property on magnesium alloy. *Mater. Lett.* **2014**, *137*, 210–213.

20. Huang, Y.; Sarkar, D.K.; Chen, X.-G. Superhydrophobic nanostructured zno thin films on aluminum alloy substrates by electrophoretic deposition process. *Appl. Surf. Sci.* **2015**, *327*, 327–334.

21. He, T.; Wang, Y.; Zhang, Y.; Lv, Q.; Xu, T.; Liu, T. Super-hydrophobic surface treatment as corrosion protection for aluminum in seawater. *Corros. Sci.* **2009**, *51*, 1757–1761.

CHAPTER 3

The Nature of Surface Oxides on Corrosion-Resistant Nickel Alloy Covered by Alkaline Water

Jiaying Cai • D. F. Gervasio

1 Department of Chemical Engineering, Arizona State University

2 Department of Chemical Engineering, University of Arizona

ABSTRACT

A nickel alloy with high chrome and molybdenum content was found to form a highly resistive and passive oxide layer. The donor density and mobility of ions in the oxide layer has been determined as a function of the electrical potential when alkaline water layers are on the alloy surface in order to account for the relative inertness of the nickel alloy in corrosive environments.

KEYWORDS

EIS Mott–Schottky; Bipolar plates; High-temperature; PEM; fuel cell; Nickel alloy

1. INTRODUCTION

Nickel metal alloys are corrosion resistant and can serve as structural materials in extraordinary environments, e.g., in long-term storage containers, high-temperature heat exchangers and aggressive chemical reactors. The stability is often attributed to the inertness of the oxides that form on the nickel alloys. One new application of these extraordinary alloys is as the structural material for a metal bipolar plate in a polymer electrolyte membrane (PEM) fuel cell stack.

The bipolar plate is among the most expensive, heaviest and voluminous components in the fuel cell stack. The bipolar plates conduct current between cells, provide flow channels for reactants and products, facilitate water and thermal management and constitute the structural backbone of a fuel cell stack. The materials for bipolar plates need to have high electric and thermal conductivity, good corrosion resistance and mechanical strength. Replacing bulky, brittle machined graphite plates by thin, durable stamped metal plates is particularly desirable for portable and mobile applications where lower bulk, fragility and cost are all needed.

After an earlier accelerated corrosion screening test [1], the high chrome molybdenum nickel alloys, such as Hastelloy C22 (composition given in Table 1), were considered one of the few materials with structural stability that is suitable for use in bipolar plates for a high-temperature PEM fuel cell stack.

Table 1. Chemical composition (wt%) of Alloy C22.

	Co	Cr	Fe	Mn	Mo	Ni	P	S	Si	V	W
C22	1.45	15.74	5.58	0.50	15.53	57.55	0.008	0.003	0.02	0.163	3.54

Compared with graphite, C22 can be made into bipolar plates from much thinner sheets. The thickness of a C22 metal sheet is <0.1 mm, whereas that of a graphite sheet is >2–5 mm. The C22 can be formed into a bipolar plate by a lower cost stamping as the manufacturing method, which costs only 10 cents to $1 per plate when compared to $5–$25 per plate for the milling or molding a graphite plate [2]. These features of metal bipolar plates are desirable for making a more compact, lighter weight and lower cost fuel cell stack.

Most importantly, alloy C22 shows remarkable corrosion resistance and stability that is suitable in the aggressive fuel cell environment. A number of studies on its general and local corrosion resistance suggest that C22 has excellent resistivity in a broad range of concentrated brines including chloride, fluoride, carbonate, sodium and calcium over a large pH and temperature range [3]. It is mainly due to the formation of a protective passive oxide layer on the surface. This occurs through electrochemical "local cell" on the metal surface, where oxygen reduction occurs at one localized metal surface site by accepting electrons generated during metal oxidation occurring at another localized site through electron conduction in the bulk metal [4]. The oxygen reduction site becomes alkaline, and metal oxidation site becomes acidic. Nickel alloy was placed in aqueous potassium hydroxide solution and exposed to various oxidizing potentials representative of a bipolar plate at an oxygen cathode. Electrochemical impedance spectroscopic (EIS) and Mott–Schottky (M–S) methods were used to determine the density and mobility of charge carriers in the passive oxide layer to understand the nature of the surface oxides and how these affect the corrosion resistance of C22 nickel alloy covered by alkaline water.

2. EXPERIMENTAL

2.1 Electrochemical Measurements

All electrochemical experiments were carried out using a three-electrode configuration at room temperature. The working electrode was nickel alloy C22 (Haynes), machined into 6 cm × 2 cm × 0.2 cm. The working electrode was abraded with 1200-grit SiC paper, polished with 1.0, 0.3 and 0.05 μm Al_2O_3 powder and then ultrasonically cleaned in deionized water. The working electrode area was 12 cm^2. A Ag/Ag_2 O reference electrode was used in 0.1 M KOH (pH 13.0) and in 1.0 M KOH (pH 13.8) electrolyte solution. The potential of the silver/silver-oxide reference electrode is 0.321 V versus RHE in 0.1 M KOH and 0.341 V versus RHE in 1.0 M KOH. This can be related to NHE (pH = 0) by the potential shift with pH using the Nernst equation. A graphite rod was used as the counter electrode. The aqueous alkaline potassium hydroxide solutions of two concentrations (0.1 M, pH 13.0 and 1.0 M, pH 13.8) were prepared using pure deionized water (PureLab Ultra system) and potassium hydroxide stock (analytical-grade reagent). The solution was deaerated with ultrapure nitrogen gas for 30 min prior to starting the experiment, and this nitrogen purge was continued throughout each experiment. Voltammetry of Alloy C22 was performed to determine the electrochemical processes that occur on the moisture-covered alloy surface. After freshly abrading the C22 working electrode, it was cathodically polarized at −1.3 V for at least 20 min to remove the air-formed oxide film, then the potential was swept from −1.3 to 0.5 V at a scan rate of 20 mV/s to survey the surface processes. The C22-alloy working electrode was held for 2 h at each film formation potential to grow the passive oxide films.

EIS and M–S tests were carried out immediately after the passive films were formed. For EIS measurements, the frequency was analyzed over a range of 10 kHz–1 MHz with a peak-to-peak modulation amplitude voltage of 20 mV. And then, the M–S experiments were done by measuring the frequency at 1 kHz during a negative potential scan from +0.2 to −1.1 V in 50 mV-increments.

All electrochemical experiments were performed using a Princeton Applied Research VMP2/Z Multichannel Potentiostat (Oak Ridge, TN) running EC-Lab version 9.13 software, and the impedance spectra analyses were performed using Zsimpwin software.

2.2 Interfacial Contact Resistance (ICR)

ICR should be minimized for bipolar plates to achieve high efficiency in PEM fuel cells. ICR measurement was conducted on the Hastelloy C22 after the electrochemical oxidization. The apparatus for measuring ICR is illustrated in Fig. 1, showing two pieces of carbon paper (SIGRACET, type GDL 10 AA, a gas diffusion layer used in PEM fuel cells) sandwiched between the sample and two copper plates. Compaction force was applied by a hydraulic press. The potential difference V across the cell and the copper plates was measured by an ohmmeter while a fixed electrical current I (0.9 A) was passed through the arrangement. The ICR was calculated as follows [5]:

$$\mathrm{ICR} = \frac{R - \mathrm{Rcp}}{2} \times A$$

where R is the total resistance (V/I), Rcp represents the resistant contribution due to the carbon paper/copper plates (~5 mΩ) and A is the sample area (cm^2). The value of ICR was greatly affected by the compaction force, and good reproducibility could be obtained only with compaction force above 200 N cm^{-2}[5, 6].

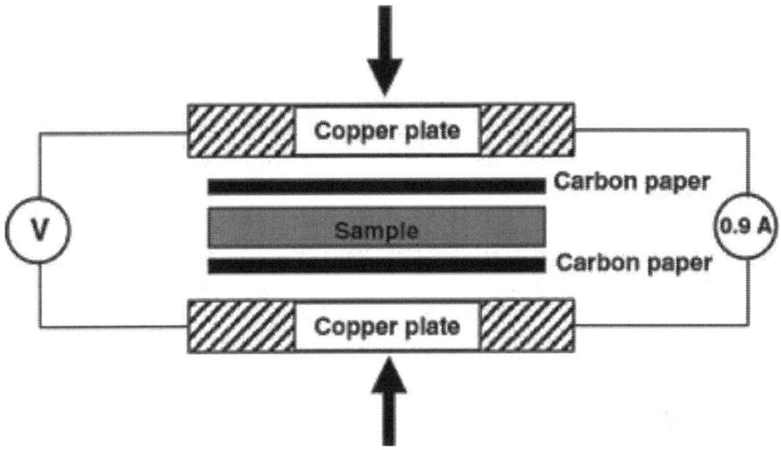

Figure 1. Apparatus used to measure interfacial contact resistance

2.3 Auger Electron Spectroscopy (AES)

In order to determine the general composition of surface-oxidized C22, AES was performed to get depth profile for oxidized samples. AES analyses were carried out on specimens at sputter rate of 2.0 nm per minute with beam current of 1.0 μA and beam voltage of 4.0 kV using Physical Electronics 590 Scanning Auger Microprobe.

3. RESULTS AND DISCUSSION

3.1 Cyclic Voltammetry

The cyclic voltammogram presented in Fig. 2 shows the surface processes occurring on alloy C22 in both 0.1 M (pH 13.0) and 1.0 M KOH (pH 13.8) solution. Figure 2a shows that the first cycle was noticeably different than the successive cycles. The first positive-going sweep shows extra anodic current from −0.7 to 0.3 V, suggesting the formation of a metal oxide layer on the alloy C22 surface. The reverse scan showed the reduction peak between 0.1 and 0.3 V in the first and succeeding negative-going scans. The second and successive positive- and negative-going scans showed growing oxidation and reduction peaks. After the third cycle, the growth rate of both oxidation and reduction peaks decreased and were virtually stable. Figure 2b shows a similar behavior for the cyclic voltammogram of the C22 in 1.0 M KOH, except there are two noticeable differences. First, there is a slight shift for the anodic peak, which was 0.3 V (vs. $Ag/Ag_2O/0.1$ M KOH) for 0.1 M KOH and 0.26 V (vs. $Ag/Ag_2O/1.0$ M KOH) for 1.0 M KOH solutions. Secondly, both the oxidation and reduction peak currents were about two times larger in the solution with 1 M versus 0.1 M KOH.

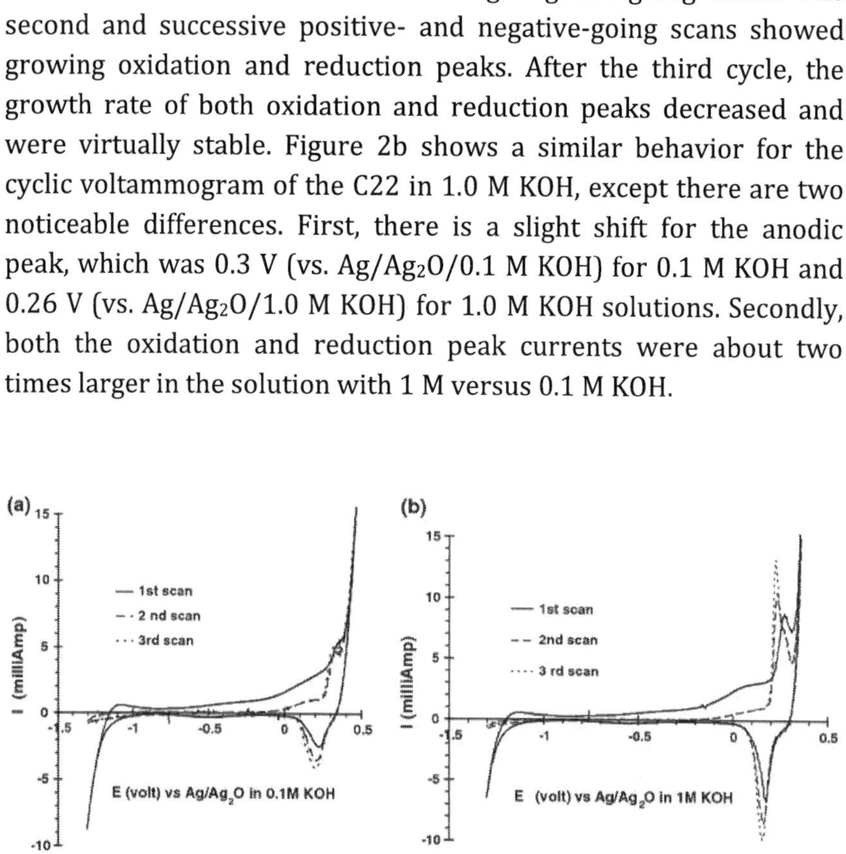

Figure 2. A, B CV of C22 in 0.1 M and 1.0 M KOH

3.2 Interfacial Contact Resistance (ICR)

Figure 3 shows the comparison of the ICR of the alloy oxidized at different potentials in both 1.0 M KOH (pH 13.8) and 0.1 M KOH (pH 13.0) solutions. The results showed that the alloy oxidized in 0.1 M

KOH had a higher ICR value than that in 1.0 M KOH solution. In both solutions, the ICR values were higher in the passive region (−0.5 to −0.1 V) and decreased at the higher potential conditions.

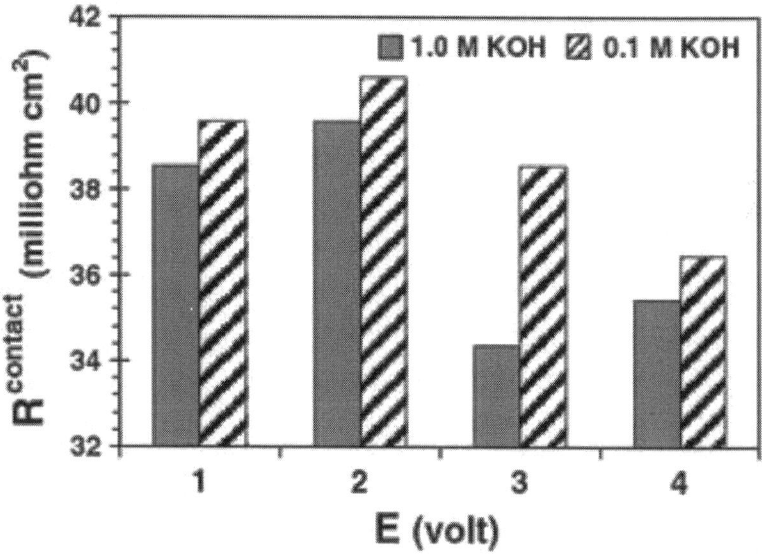

Figure 3. Interfacial contact resistance of alloy C22 after oxidized at −0.5, −0.1, 0.1 and 0.2 V in 1.0 and 0.1-M KOH solutions

Generally, the influence of Cr-oxide on the Ni-based material resistance is very complex, and it can be considered that the decrease of conductivity follows the trend that the conductivity of Ni-oxide is greater than the conductivity of Cr-oxide [5]. Therefore, it appears that when alloy C22 is oxidized in 0.1 M KOH solution, a larger amount of Cr-oxide forms on the surface, which results in a higher value of ICR. The depth profile for the oxide films on C22 by AES (not shown here) showed more Cr-oxide was formed in 0.1 M KOH, which is consistent with this assertion.

3.3 Impedance Measurement

EIS and M–S tests were carried out on the passive films formed at different potentials in order to investigate the influence of the film formation potential on the character of passive films on alloy C22. The Nyquist plots are shown in Fig. 4a and 4c for the nickel alloy in 1.0 and 0.1 M KOH electrolyte. The impedance data can be modeled by a simple equivalent circuit Rs (CscRp), where Rs is the electrolyte solution resistance, *C* sc is the space charge capacity and Rp is the polarization resistance. It is clear that the impedance response is sensitive to the film formation potential. In both 0.1- and 1-M KOH solutions, smaller arcs were observed in the potential range of 0.2 and 0.4 V, while larger ascending arcs, which do not form semicircles on the real axis, are observed between −0.3 and −0.1 V. This phenomenon is more clearly shown in Fig. 4b and 4d, where Rp initially increased with potentials (within the passive range), but when potentials are within the trans-passive region ($E > -0.1$ V), Rp decreases with *E*. The existence of the resistance Rp versus *E* peak can be attributed to the establishment of passive oxide layer in the beginning and then the oxidative ejection of chromium cations from the barrier oxide layer [7].

The impedance behavior for alloy C22 in the 0.1- and 1-M KOH solutions show one systematic difference, namely, the arcs are always larger in 0.1 M KOH. It appears that the higher concentration of $[OH]^-$ ions results in a less-resistive passive oxide film on the nickel alloy surface, especially in the potential range between −0.5 and −0.1 V. The possible formation process of metal oxide is presented as follows.

$$M \rightarrow M^{x+} + e^{x-}$$

$$[OH]^- + M^{x+} \rightarrow M[OH]_x \rightarrow MO_{x/2}$$

Having more $[OH]^-$ ions in solution favors the above reaction, and hence, the quick formation of an passive oxide layer, which covers the metal surface and slowed down the further oxidization of metal.

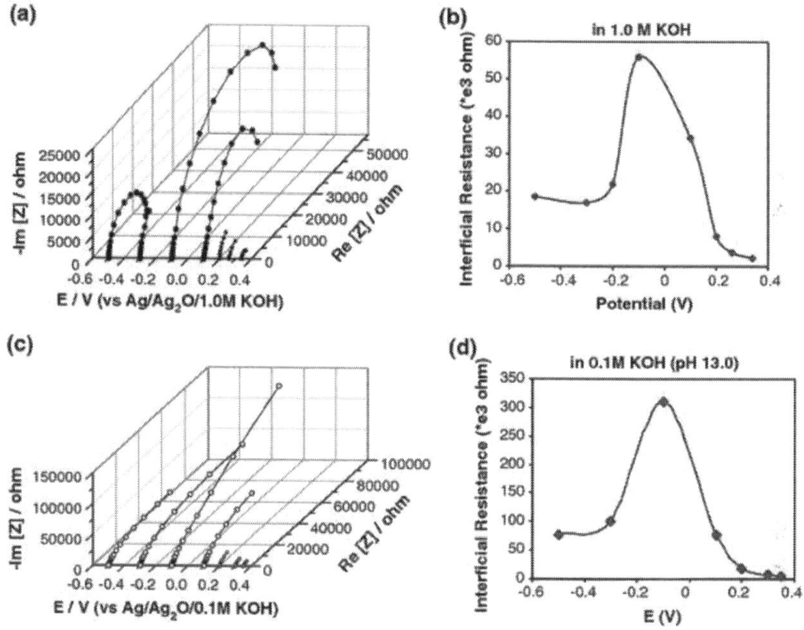

Figure 4. A, B EIS of C22 in 1.0 M KOH. **C, D** EIS of C22 in 0.1 M KOH

Following each EIS measurement, an M–S test was performed to study the semiconducting properties of a passive oxide film that was formed on the surface of the nickel alloy. The M–S analysis measures the electrode capacitance as a function of potential. Under depletion conditions, the M–S relationship is given by Eq. (1)

$$\frac{1}{C_{SC}^2} = \frac{2}{\epsilon\varepsilon\varepsilon_0 N A^2}\left(V_E - V_{fb} - \frac{kT}{\epsilon}\right) \quad (1)$$

where C_{SC} is the space charge capacitance, ε is the dielectric constant of the semiconductor, ε_0 is permittivity of free space ($8.854e^{-14}$ F/cm), N is defect density (electron donor concentration for n-type semiconductor or hole acceptor concentration for p-type semiconductor) and k is the Boltzmann constant. kT/e is the thermal voltage, which is the voltage a single charge falls through to pick up the thermal energy. kT/e is about 25 mV at the ambient temperature.

The M–S analysis assumes the space charge capacitance is much smaller than the double-layer capacitance such that the contribution of double-layer capacitance to the total capacitance value could be negligible. For a p-type semiconductor, C_{SC}^{-2} versus E should be linear with a negative slope, which is inversely proportional to the acceptor density N. For an n-type semiconductor, the slope should be positive.

Figure 5 shows the M–S plots recorded at 1 kHz frequency for passive films formed on Alloy C22 in 1.0- and 0.1-M KOH solutions at different potentials.

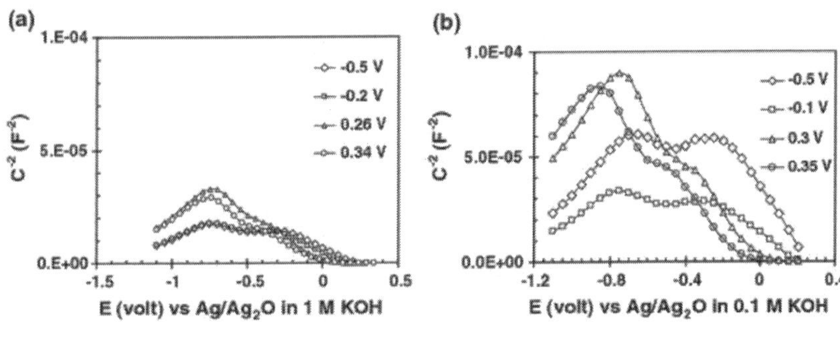

Figure 5. A, B M–S test of C22 in 1.0 and 0.1 M KOH

As shown in Fig. 5b, the capacitance decreased (C_{SC}^{-2} increased) at low potentials ($-1.1 < E < -0.8$ V), suggesting an n-type semiconductor. At higher potentials ($E > -0.1$ V), however, the capacitance increased (C_{SC}^{-2} decreased), showing a p-type semiconductor. The change of the electronic character is more likely due to the generation of the cation vacancies at film/solution interface through the oxidative ejection of cations from the film [8]. This result is consistent with the above Nyquist plots where the most resistant film was formed at the potential of −0.1 V, where the change of electronic character appeared.

Over the potential range between −0.8 and −0.1 V, the capacitance was nearly constant, for those passive films formed at lower potentials (−0.5, −0.3, −0.2, −0.1 and 0.1 V). This phenomenon was also reported by Da Belo et al. [9] on Ni-20% Cr alloy in pH 9.2 borate buffer. For those passive films formed at higher potentials (0.2, 0.26 and 0.34 V), there was no clear potential range over which the capacitance varies slightly. Their M–S profiles behaved similar to those of the films on pure Cr, which presents a peak in the C_{SC}^{-2} versus E plots followed by a steadily linear region negative slope (see [10]).

Defect density N of the passive films could also be determined by the slope of the linear part of M–S profile. Both the donor density calculated from the n-type part and the acceptor density from the p-type part in passive films formed in 1.0-M KOH electrolyte solution are larger than those formed in 0.1-M KOH solution (see Fig. 6). The higher defects concentration within the film resulted in lower resistant passive films, and accordingly, higher conductivity, which was in a good agreement with the ICR and Nyquist results.

3.4 AES Depth Profile

Figure 7 showed the content of three major components within the surface oxide films on alloy C22 versus the depth of the films.

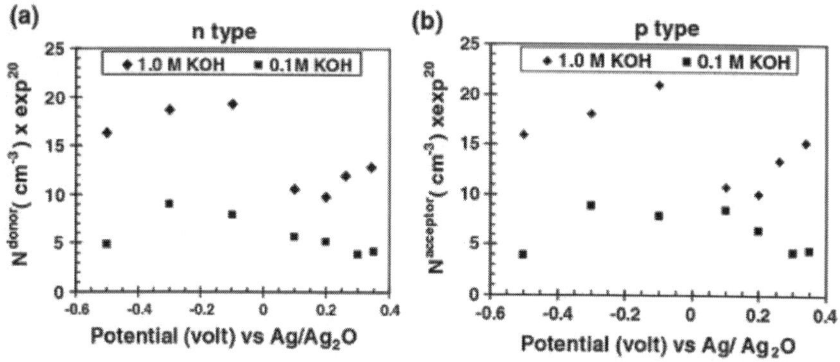

Figure 6. A, B Donor density (acceptor density) versus film formation potentials

In all cases, the amount of Cr-oxide was slightly higher in 0.1-M KOH than that in 1.0-M KOH solution. This result is consistent with the effect of solution pH on the ICR value, which was higher for the oxide films formed in 0.1 M KOH.

The depth profile (*b*) behaved quite different from the other two cases. For the oxide film formed at −0.1 V, the content of Cr-oxide is higher in the outer layer of the film, which was ~51% formed in 1.0 M KOH and ~55% in 0.1 M KOH compared with ~20% in the bulk alloy. It decreased greatly from the outer to inner surface at the depth of 2 nm, while the content of Ni-oxide increased and finally dominated in the inner layer of the film. However, for the oxide films formed at −0.5 and 0.26 V, this dual-layered structure was not observed. And the Ni-oxides were dominant through the entire oxide film. This result could also be explained the highest value of ICR for the oxide film formed at −0.1 V, which the higher amount of Cr-oxide was responsible for the higher contact resistance. The thickness of the oxide films was estimated by the depth profile at the range of 3–4 nm, where the three components Cr, Ni and Mo converged to a state value, respectively.

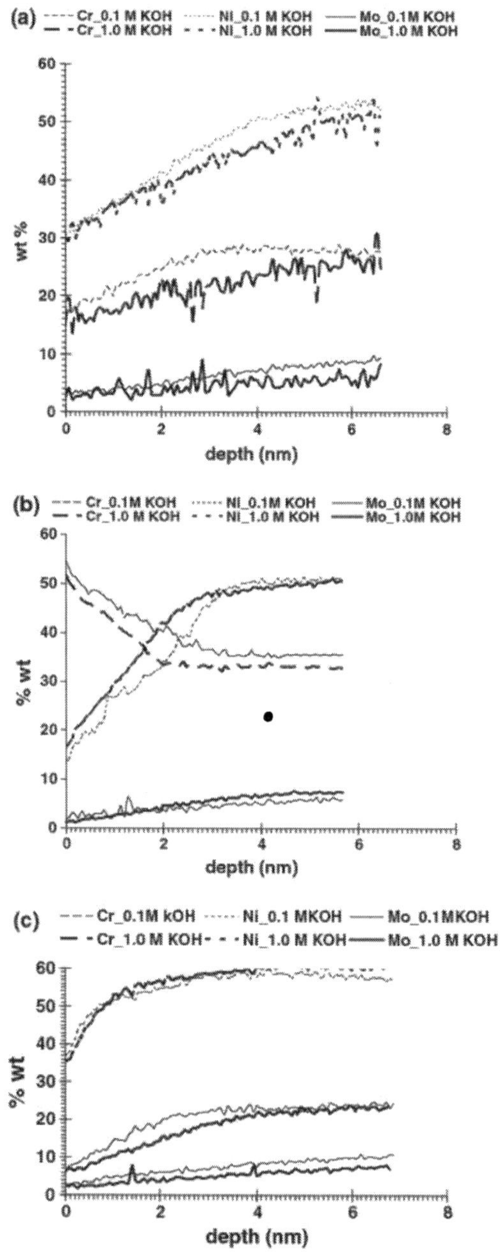

Figure 7. AES depth profile of the oxide film on alloy C22 formed at −0.5 V (**A**), −0.1 V (**B**) and 0.26 V (**C**) in 0.1 and 1.0 M KOH

4. CONCLUSIONS

The oxide film that forms on nickel alloy C22 is affected by film formation potential and pH. ICR and EIS show the interfacial film resistance Rp is sensitive to the film formation potential. The current for the formation of oxide peaks at the potential of −0.1 V. More concentrated KOH electrolyte solution contributes to the formation of less resistant and hence larger peak current for this passive film formation at 0.1 V on nickel alloy C22. The M–S analysis of the oxide layer on nickel alloy C22 shows that the oxide film on the nickel alloy is semiconducting when formed in both 0.1- and 1-M KOH solutions. Over lower potential range, the oxide film on nickel alloy C22 displays n-type character, while p-type character is found at higher potentials. Defect concentration obtained from the M–S plots is higher when the film is formed in 1.0-M KOH solution at all the investigated film formation potentials, which is consistent with a lower film resistance for oxides formed in 1-M compared to 0.1-M KOH solution. The AES depth profile shows a dual-layered structure in the oxide film formed at −0.1 V, where a Cr-rich outer layer is responsible for the higher contact resistance. The amount of Cr showed in the depth profile was higher in 0.1-M KOH than that in 1.0-M KOH solution, which further confirmed that a more resistive oxide film grows on the nickel alloy when it is covered by a less concentrated aqueous KOH (less basic) solution.

As is, this alloy is stable enough to be used as a bipolar plate in a high-temperature polymer electrolyte membrane fuel cell (HT PEM FC), but the surface conductance of this alloy is too low to be used as a bipolar. However, coating with a thin stable conductive layer, such as gold, will give suitable surface conductivity. Because the Hastelloy C22 is inert to corrosion, defect in the gold coating will not grow, and a gold-coated Hastelloy C22 bipolar plate should be suitable for use in a HT PEM fuel cell. Ongoing work concerns testing this assertion in HT PEM fuel cell stacks.

REFERENCES

1. Gervasio D, Kinder J, Hoskins N, Onyeabor V, Taghavi S, Pattekar AV, Fleckner K: ***COST-BENEFIT ANALYSIS OF COMMERCIAL BIPOLAR PLATES FOR PEMFC'S.*** In Naval Air Systems Command. Nathan Kumbar, NAVAIR, New Orleans; 2006. High temperature polymer electrolyte membrane fuel cells (HT-PEMFCs) for portable power, Paper number 2006–01–3096, Power Systems Conference, Fuel Cells (Session Code: PSC6), Organizers—Michael Allen High temperature polymer electrolyte membrane fuel cells (HT-PEMFCs) for portable power, Paper number 2006-01-3096, Power Systems Conference, Fuel Cells (Session Code: PSC6), Organizers—Michael Allen

2. Jayakumer K, Pandiyan S, Rajakshmi N, Dhathathreyan KS: ***COST-BENEFIT ANALYSIS OF COMMERCIAL BIPOLAR PLATES FOR PEMFC'S.*** J. Power Sources 2006, ***161:*** 454–459. 10.1016/j.jpowsour.2006.04.128

3. Gray JJ, Orme CA: ***ELECTROCHEMICAL IMPEDANCE SPECTROSCOPY STUDY OF THE PASSIVE FILMS OF ALLOY 22 IN LOW PH NITRATE AND CHLORIDE ENVIRONMENTS.*** Electrochim. Acta 2007, ***52:*** 2370–2375. COI number [1:CAS:528:DC%2BD2sXpslWksQ%3D%3D] COI number [1:CAS:528:DC%2BD2sXpslWksQ%3D%3D] 10.1016/j.electacta.2006.08.043

4. Rogers SP, Gervasio DF: ***OXYGEN ELECTRO-REDUCTION ON C22 AND C276 NICKEL METAL ALLOY.*** Electrochem. Soc. Trans. 2006, ***3:*** 431–442.

5. Silva RF, Franchi D, Leone A, Pilloni L, Masci A, Pozio A: ***SURFACE CONDUCTIVITY AND STABILITY OF METALLIC BIPOLAR PLATE MATERIALS FOR POLYMER ELECTROLYTE FUEL CELLS.*** Electrochim. Acta 2006, ***51:*** 3592–3598. COI number

[1:CAS:528:DC%2BD28XjsFeht7s%3D] COI number [1:CAS: 528:DC%2BD28XjsFeht7s%3D] 10.1016/j.electacta.2005.10.015

6. Pozi A, Zaza F, Masci A, Silva RF: ***BIPOLAR PLATE MATERIALS FOR PEMFCS: A CONDUCTIVITY AND STABILITY STUDY.*** J. Power Sources 2008, ***179:*** 631–639. 10.1016/j.jpowsour.2008. 01.038

7. Priyantha N, Jayaweera P, MacDonald DD, Sun A: ***AN ELECTROCHEMICAL IMPEDANCE STUDY OF ALLOY 22 IN NACL BRINE AT ELEVATED TEMPERATURE. I. CORROSION BEHAVIOR.*** J. Electroanal. Chem. 2004, ***572:*** 409–419. COI number [1:CAS:528:DC%2BD2cXotV2gsr0%3D] COI number [1: CAS:528:DC%2BD2cXotV2gsr0%3D] 10.1016/j.jelechem.2004. 06.031

8. Macdonald DD, Sun A, Priyantha N, Jayaweera P: ***AN ELECTROCHEMICAL IMPEDANCE STUDY OF ALLOY-22 IN NACL BRINE AT ELEVATED TEMPERATURE: II. REACTION MECHANISM ANALYSIS.*** J. Electrochem. Chem. 2004, ***572:*** 421–431. COI number [1:CAS:528:DC%2BD2cXotV2gsro%3D] COI number [1:CAS:528:DC%2BD2cXotV2gsro%3D] 10.1016/j. jelechem.2004.05.032

9. Da Belo M, Hakiki NE, Ferreira MGS: ***SEMICONDUCTING PROPERTIES OF PASSIVE FILMS FORMED ON NICKEL-BASE ALLOYS TYPE ALLOY 600: INFLUENCE OF THE ALLOYING ELEMENTS.*** Electrochim. Acta 1999, ***44:*** 2473–2481. 10.1016/ S0013-4686(98)00372-7

10. Bojinov M, Fabricius G, Kinnunen P, Laitinen T, Mäkelä K, Saario T, Sundholm G: ***ELECTROCHEMICAL STUDY OF THE PASSIVE BEHAVIOUR OF NI-CR ALLOYS IN A BORATE SOLUTION-A MIXED-CONDUCTION MODEL APPROACH.*** J. Electroanal. Chem. 2001, ***504:*** 29–44. COI number [1:CAS:528: DC%2BD3

MXjtV2ms7Y%3D] COI number [1:CAS: 528:DC%2BD3MXjtV2ms7Y%3D] 10.1016/S0022-0728(01) 00423-5

CHAPTER 4

High Temperature Oxidation and Corrosion Properties of High Entropy Superalloys

Te-Kang Tsao [1], An-Chou Yeh [1,*], Chen-Ming Kuo [2] and Hideyuki Murakami [3]

[1] Department of Materials Science and Engineering, National Tsing Hua University, Hsinchu 30013, Taiwan

[2] Department of Mechanical and Automation Engineering, I-Shou University, Kaohsiung 84001, Taiwan

[3] National Institute for Materials Science, Sengen 1-2-1, Tsukuba, Ibaraki 305-0047, Japan

ABSTRACT

The present work investigates the high temperature oxidation and corrosion behaviour of high entropy superalloys (HESA). A high content of various solutes in HESA leads to formation of complex oxides, however the Cr and Al activities of HESA are sufficient to promote protective chromia or alumina formation on the surface. By comparing the oxidation and corrosion resistances of a Ni-based superalloy—CM247LC, Al2O3-forming HESA can possess comparable oxidation resistance at 1100 °C, and Cr2O3-forming HESA can exhibit superior resistance against hot corrosion at 900 °C. This work has demonstrated the potential of HESA to maintain surface stability in oxidizing and corrosive environments.

KEYWORDS

alloy; superalloys; oxidation; hot corrosion

1. INTRODUCTION

Many gas turbine engine components are made of Ni-based superalloys, and the continuing strive to improve the engine efficiency has demanded improved temperature capability in these materials. As a result, high levels of refractory elements have been added to Ni-based superalloys to improve their high temperature strength [1,2,3,4,5]. Nevertheless, this has led to high alloy density and a high propensity to form refractory oxides that degrade the surface stability [6,7,8,9,10]. The ability to maintain surface stability in oxidizing and corrosive environments is one of the most critical requirements for high temperature application materials, since material loss and surface degradation due to oxidation and corrosion can ultimately lead to the failure of structural components [11,12,13,14]. Recently, novel high temperature alloys based on the "high entropy alloy" concept have been designed by incorporating both sluggish diffusion and lattice distortion strengthening effects [15,16,17,18], and these materials have been referred to as "high entropy superalloys" (HESAs) [19]; this alloy design approach allows HESAs to be strengthened by high contents of various solutes rather than alloying with a high content of refractory elements. HESAs possess similar microstructures to those of Ni-based superalloys, which is thermodynamically stable FCC γ and L12 γ′, and our previous study has shown that HESAs can exhibit higher hardness at elevated temperature than conventional Ni-based superalloys. Furthermore, the measured alloy densities of HESA are below 8.0 g/cm3, which are relatively lower than that of conventional Ni-based superalloys (8.5–9.0 g/cm3), and the raw material cost of HESAs can be 20% less than that of CM247LC [19], so they have the potential to become more cost-effective. So far the surface stability of HESAs has not yet been reported. The aim of present work was to investigate the high temperature oxidation and corrosion behaviour of HESAs, and their potential for high temperature applications with respect to surface stability will be discussed in this article.

2. MATERIAL AND METHODS

The alloys of interest are listed in Table 1, where the two high entropy superalloys are designated as HESA-1 and HESA-2. Both HESAs are based on a Ni-Co-Fe-Al-Cr-Ti system. The Al content in HESA-2 is slightly higher than that of HESA-1, and HESA-2 contains minor refractory additives. Their densities are relatively low compared to that of most conventional Ni-based superalloys [5,6]. Master alloys were prepared by vacuum-arc-melting, and then casting ingots with designed compositions were obtained by vacuum-induction-melting process. The Bridgman method was then applied to cast samples by directional solidification (DS). The withdraw process started after the mold temperature was stable at 1550 °C, and the withdraw rate was 40 mm·h−1 with a temperature gradient of about 30 °C·mm−1; the setup of the DS casting has been described in our previous work [20]. From previous measurements by differential scanning calorimetry (DSC), the γ′ solvus temperatures of HESA-1 and HESA-2 are 1165 °C and 1194 °C, respectively. Therefore, the DS HESA samples (10 mm in diameter, and 120 mm in length) were solution-heat-treated in a vacuum chamber at 1210 °C for 10 h to resolve γ′ particles and homogenize alloying elements [19]. For the following ageing process to uniformly reprecipitate γ′ particles, different ageing temperatures from 800 to 1100 °C with different ageing times were tested, and 1000 °C/3 h was determined to be the primary ageing condition due to preferred γ′ volume fraction (between 50% and 70% for HESA-1 and HESA-2) and also size of γ′ (around 300 nm for both HESAs). The secondary ageing can further adjust γ′ morphology to be more regular, asnd it was conducted at 800 °C for 20 h. Samples of a Ni-based superalloy—CM247LC processed by the same DS casting process and standard heat treatment [21,22] were examined for comparative studies.

Table 1. The nominal composition, calculated mixing entropy (ΔSmix) and alloy density of HESA-1, HESA-2 and CM247LC [19,21].

	Composition (wt.%)										ΔS_{mix}	Density (g/cm^3)
	Ni	Al	Co	Cr	Fe	Ti	Hf	Ta	Mo	W		
HESA-1	44.0	3.9	22.3	11.7	11.8	6.3	–	–	–	–	1.58R	7.64
HESA-2	51.0	5.0	18.0	7.0	9.0	5.0	–	2.0	1.5	1.5	1.56R	7.94
CM247LC	61.8	5.6	9.2	8.1	–	0.7	1.4	3.2	0.5	9.5	1.29R	8.50

Specimens for oxidation and hot corrosion tests were machined to a dimension of 8 mm × 8 mm × 3 mm by electrical discharge machining. All surfaces were ground by 800-grit SiC paper and ultrasonically cleaned. Isothermal oxidation tests were carried out at 900 and 1100 °C for 5, 20, 50, 100 and 200 h. Each measured weight change from the corresponding test temperature and time was obtained with a different specimen. The specimen was put into a 3 cm tall ceramic container, heated in still air inside a box furnace, and removed from the furnace to cool. No obvious oxide spallation was observed after high temperature exposure. The weight of each specimen was then measured by an electronic weight balance, and the weight gain was estimated by subtracting the weight of the initial sample from the final weight of the oxidized sample. The hot corrosion tests were conducted by both salt-coated and crucible tests. For the salt-coated method, the same specimen was utilized for each alloy. The top surface of specimen was coated uniformly with 75% Na_2SO_4 + 25% NaCl salts (0.2 kg/m^2) and water was evaporated by putting the specimen on a hot plate heated up to 200 °C prior to the 900 °C exposure. After 20 h exposure, the specimen was removed from the furnace, cooled to room temperature and carefully washed with hot distilled water. The weight change was measured afterwards. Then, the specimen was recoated with a similar amount of salts, heated at 900 °C for another 20 h, washed with hot distilled water and again the weight loss for the second cycle measured until the accumulation of a total of 100 h of testing time. For the crucible test, a specimen of each alloy was immersed entirely into a 75% Na_2SO_4 + 25% NaCl solution. After exposure at 900 °C for 20 h, the specimen was air cooled to room temperature and also carefully washed with hot distilled water; the average length change of the specimen dimensions was then recorded.

The metallographic specimens were prepared by the process of mounting, grinding and polishing. Scanning electron microscopes (SEM, S-4700, Hitachi, Tokyo, Japan and JEOL-5410, Jeol, Akishima, Japan) equipped with energy dispersive X-ray spectrometry (EDS) were used to observe and analyze the microstructure and oxide scales. The oxides were identified by an X-ray diffractometer (XRD 6000, Shimadzu, Kyoto, Japan) with Cu-target radiation at 30 kV and 20 mA. The specimens were scanned at 2θ angles from 20 to 100° with a scanning rate of 2 $deg \cdot min^{-1}$. XRD spectra were analyzed by search-match software based on the JCPDS database [23].

3. RESULTS

3.1. Oxidation Behaviour

The isothermal oxidation behaviours of HESA-1, HESA-2 and CM247LC at 900 °C and 1100 °C are presented in Figure 1a,b, respectively. The oxidation weight gain of HESA-1 is the highest, followed by that of HESA-2 and CM247LC at both temperatures. Furthermore, the oxidation behaviour of HESA-2 appears to be similar to that of CM247LC, especially at the high temperature of 1100 °C.

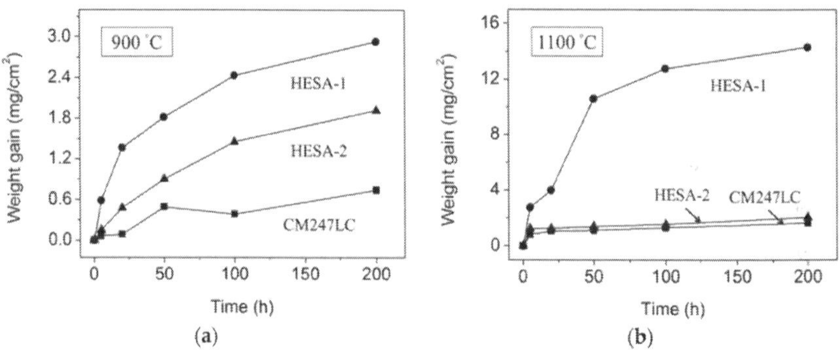

Figure 1. The isothermal oxidation behaviours of HESA-1, HESA-2 and CM247LC at (a) 900 °C; (b) 1100 °C.

Figure 2 shows the cross-sectional oxide scales and microstructures of HESA-1 and HESA-2 after 5 h exposure at 900 °C and 1100 °C. A γ' depletion zone can be observed between the oxide layer and the matrix due to loss of Al during oxidation, and the thickness of γ' depletion zone for HESA-1 appears to be larger than that of HESA-2. Types of oxide have been characterized by the contrast from backscattering images, SEM-EDS measurements and XRD analyses.

The measured oxide concentrations should approximately agree with the compositions indexed by XRD. The XRD analysis are shown in Figure 3, and it shows that identical type of oxides are formed at 900 °C and 1100 °C, Figure 3a,b. The oxides identified on CM247LC include NiO, $NiCr_2O_4$, $NiAl_2O_4$, HfO_2 and Al_2O_3 that agree with previous studies [24,25]. The oxide scales on HESA-1 are (Ni, Co)O, $CoFe_2O_4$ and $Fe_3Ti_3O_{10}$, Cr_2O_3 followed by TiO_2 and Al_2O_3. As for HESA-2, (Ni, Co)O, $CoFe_2O_4$, $(Ni, Ti)_3O_4$, $CrTi_2O_5$, $CrTaO_4$ and Al_2O_3 are shown. From Figure 2a,b, at 900 °C, neither Cr_2O_3 nor Al_2O_3 are formed continuously after 5 h. As for the 1100 °C/5 h exposure results, continuous Al_2O_3 can be observed in HESA-2 (Figure 2d), while thick external (Ni, Co)O, Co, Fe, Ni and Ti-rich oxides, Cr_2O_3 and discontinuous Al_2O_3 are present in HESA-1 (Figure 2c). Although both Cr_2O_3 and Al_2O_3 can provide surface protection, Al_2O_3 would possess lower oxygen permeability and be more thermodynamically stable than Cr_2O_3, while Cr_2O_3 may gradually transform into the volatile CrO_3 beyond 950 °C ($2Cr_2O_3$ (s) + $3O_2$ (g) = $4CrO_3$ (g)) [26,27,28,29,30]. Hence, the formation of a continuous Al_2O_3 layer is critical for protection against oxidation at high temperature.

Figure 4 shows the oxide scales of both HESA alloys after oxidizing at 900 °C for 100 h, and this can be an example for the identification of oxides. From Figure 4a,b, the outer layers of HESA-1 and HESA-2 are relatively bright in the SEM-BSE images, and the measured compositions are rich in Ni, Co and O, Figure 4e,f, so they can be identified as (Ni, Co)O which corresponds with the XRD scanning results. The same methods were applied to all oxide regions, and the oxides on both alloys were then determined. The SEM-EDS mapping results also agree with the oxide distribution, Figure 4c,d. The outer scales of both alloys are mainly composed of (Ni, Co)O and oxides rich in Co, Fe, Ni and Ti. Moreover, HESA-1 possesses a thick Cr_2O_3 layer with discontinuous Al_2O_3 beneath it (Figure 4a,c). By contrast, HESA-2 contains some $CrTaO_4$ with a sublayer of continuous Al_2O_3

(Figure 4b,d). The cross-sectional oxide scales of HESA-1 and HESA-2 after 100 h at 1100 °C are shown in Figure 5. The upper part of the SEM images in Figure 5a,b is caused by measurement artifacts due to the signals of mounting region. SEM-EDS mapping also proves that HESA-1 exhibits a layer of Cr2O3, and there is discontinuous Al2O3 underneath it (Figure 5a,c).

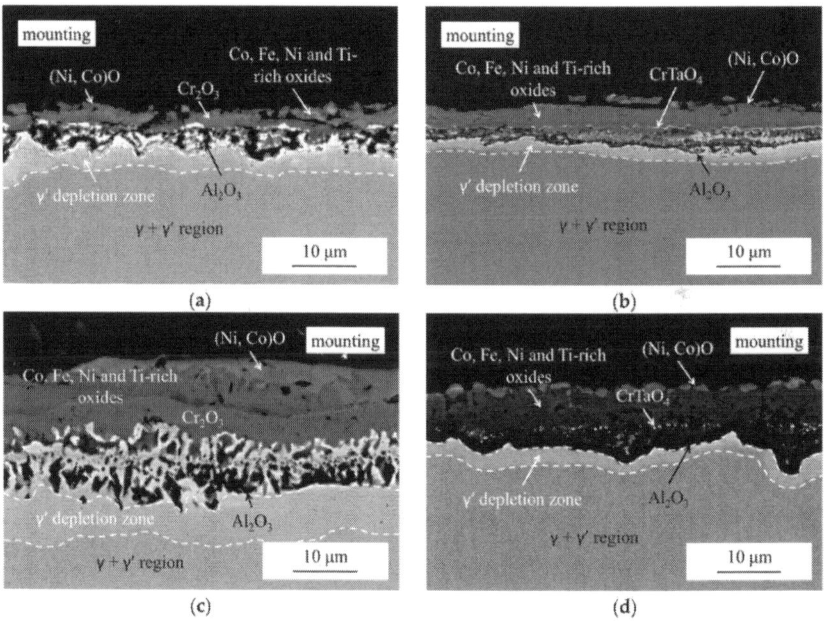

Figure 2. SEM-BSE images of samples after oxidation, (a) HESA-1 after 900 °C/5 h; (b) HESA-2 after 900 °C/5 h; (c) HESA-1 after 1100 °C/5 h; (d) HESA-2 after 1100 °C/5 h.

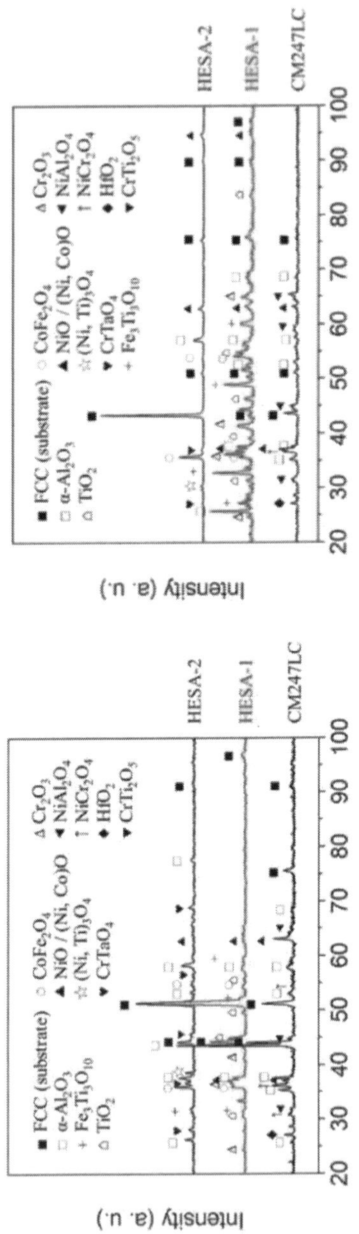

Figure 3. XRD analysis of oxidized HESA-1, HESA-2 and CM247LC after (a) 900 °C/5 h; (b) 1100 °C/5 h exposure.

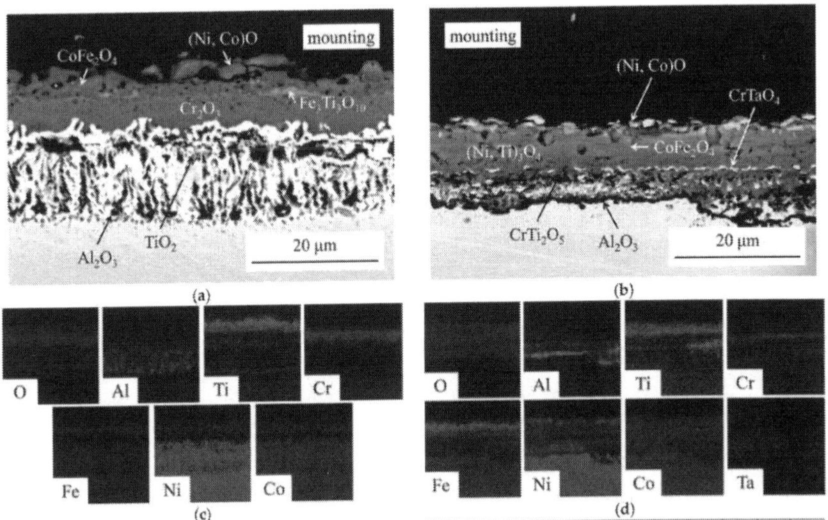

at%	Ni	Al	Co	Cr	Fe	Ti	O
(Ni, Co)O	30.6	1.8	11.6	2.1	6.1	1.6	46.2
$CoFe_2O_4$	4.4	0.3	11.4	0.6	18.5	5.2	59.6
$Fe_3Ti_3O_{10}$	0.9	0.6	3.2	2.4	15.0	16.7	61.2
Cr_2O_3	0.4	2.1	0.3	35.9	0.4	1.2	59.7
TiO_2	4.9	0.5	3.1	2.1	4.9	23.3	61.2
Al_2O_3	2.6	35.0	1.3	1.5	0.5	2.1	57.0

(e)

at%	Ni	Al	Co	Cr	Fe	Ti	Ta	Mo	W	O
(Ni, Co)O	36.3	--	7.3	0.3	2.7	0.4	0.1	0.2	0.1	52.6
$CoFe_2O_4$	8.3	2.1	11.4	2.8	18.2	6.4	0.1	--	--	50.7
$(Ni, Ti)_3O_4$	16.6	4.6	5.4	4.7	3.8	20.0	0.5	0.2	1.2	43.0
$CrTi_2O_5$	2.4	7.2	1.2	9.3	1.7	16.9	4.5	0.1	0.1	56.6
$CrTaO_4$	1.8	1.8	1.2	14.3	1.3	7.0	15.7	--	2.3	54.6
Al_2O_3	0.3	40.6	0.1	0.2	0.1	0.1	--	--	--	58.6

(f)

Figure 4. Oxidized microstructures of (a) HESA-1 and (b) HESA-2 after 900 °C/100 h exposure. SEM-EDS mapping of (c) HESA-1 and (d) HESA-2. The measured composition of oxides on (e) HESA-1 and (f) HESA-2.

For HESA-2, continuous Al2O3 can form to resist internal oxidation (Figure 5b,d). Comparing to that of 5 h exposure, Figure 2c,d, it is apparent that further internal oxidation has occurred after 100 h for HESA-1 with Cr2O3 protection, while that of HESA-2 can be effectively hindered by the continuous Al2O3 layer. This corresponds with the more obvious oxidation weight gain of HESA-1, Figure 1b. As a result, the oxidation resistance of HESA-2 can be superior to that of HESA-1. Figure 6 shows the oxide scales of CM247LC after 900 °C and 1100 °C oxidation. After 5 h exposure, its outer oxides consist of NiO and spinels; although compared to that of HESA-1 and HESA-2, its Al2O3 has not formed so continuously (Figure 6a,b), CM247LC does not contain complex oxides such as TiO2, CrTi2O5, and CrTaO4, so its oxidation weight gain can still be the least among three alloys. The Al2O3 forms continuously at both temperatures after 100 h exposure for CM247LC (Figure 6c,d), like in the case of HESA-2.

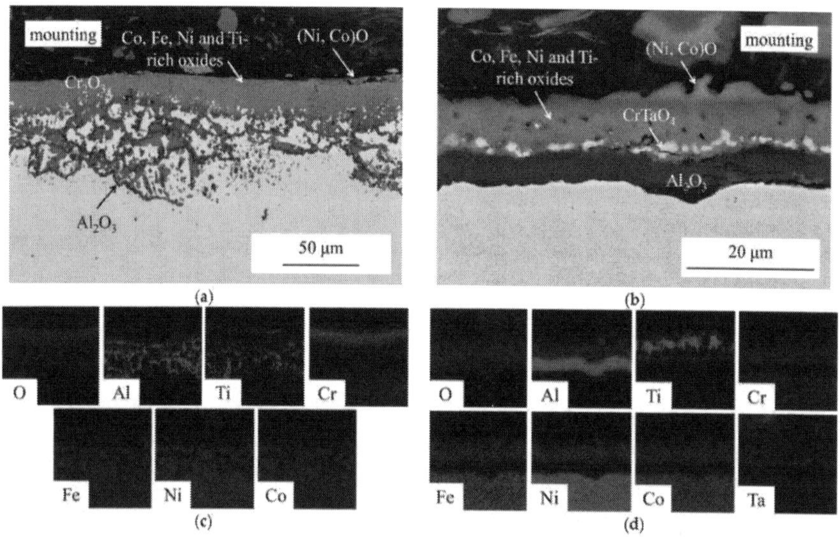

Figure 5. Oxidized microstructures of (a) HESA-1 and (b) HESA-2 after 1100 °C/100 h exposure. SEM-EDS mapping of (c) HESA-1; and (d) HESA-2.

3.2. Hot Corrosion Behaviour

Hot corrosion is an accelerated corrosion process induced by the formation of solid and molten salts at high temperature. The mechanism involves two main steps. Initially, the fluxing of corroding salts would attack the protective oxides, making the substrate to be in contact with the salts and suffer from internal sulfidation, which causes metal losses during the second step [31]. At temperatures above the melting point of the predominant salt deposit Na2SO4, basic fluxing of Na2SO4 with the protective oxide layer would occur as follows [32]:

$$Na_2SO_4 \Rightarrow Na_2O + SO_3$$

$$2Al_2O_3 + 2Na_2O + O_2 \Rightarrow 4NaAlO_2 \text{ (for } Al_2O_3\text{) and } 2Cr_2O_3 + 2Na_2O + O_2 \Rightarrow 4NaCrO_2 \text{ (for } Cr_2O_3\text{)}.$$

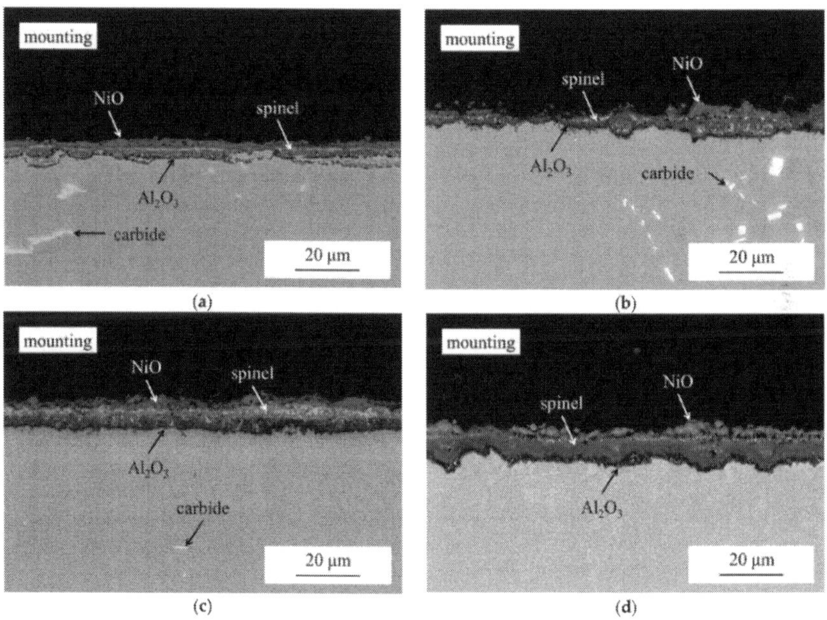

Figure 6. Oxidized microstructures of CM247LC after (a) 900 °C/5 h; (b) 1100 °C/5 h; (c) 900 °C/100 h; and (d) 1100 °C/100 h exposure.

The presence of NaCl may form eutectic mixtures with Na2SO4 and a further decrease in its melting point, resulting in more severe corrosion due to molten salts. The mass changes as a function of corrosion time for HESA-1, HESA-2 and CM247LC after 900 °C salt-coated tests are shown in Figure 7a. The same salt solution was utilized for coating, and with the same specimen size of top surface, and the amounts of salts were controlled equally for each cycle and alloy. During the salt-coated test, corrosion would occur on the top surface while oxidation weight gain appeared on the sides. After the initial 20 h test, the combination of weight loss on the top surface and weight gain on the sides are similar among three alloys, and weight changes become more apparent from the second salt-coated cycle onwards. HESA-1 exhibits the least weight loss up to 100 h, compared to HESA-2 and CM24LC. Figure 7b presents the average sample dimension changes after 900 °C/20 h crucible tests. The surfaces were all washed carefully with hot distilled water. HESA-1 also exhibits less dimensional loss comparing to HESA-2 and CM247LC, and the corrosion behaviour of HESA-2 resembles that of CM247LC.

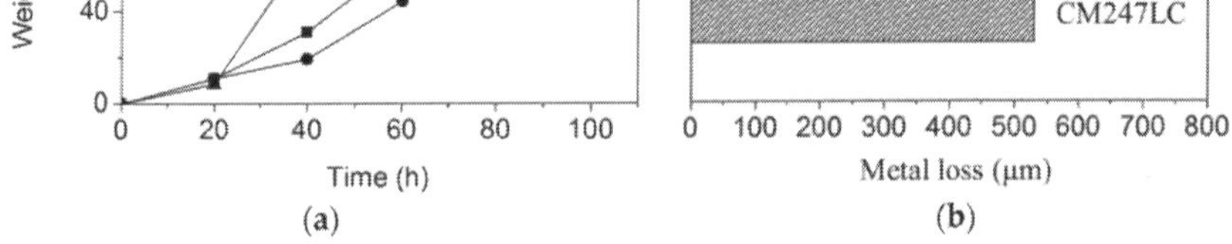

Figure 7. The corrosion behaviours of HESA-1, HESA-2 and CM247LC at 900 °C (a) weight loss of salt-coated tests; and (b) average specimen dimension loss of crucible tests.

During hot corrosion, sulfides would form at the interface between the oxide and substrate, and with longer reaction times, it has been reported that these sulfides may be oxidized and further react at the oxide front [33]. Figure 8 shows the microstructure observations after 900 °C salt-coated tests and crucible tests.

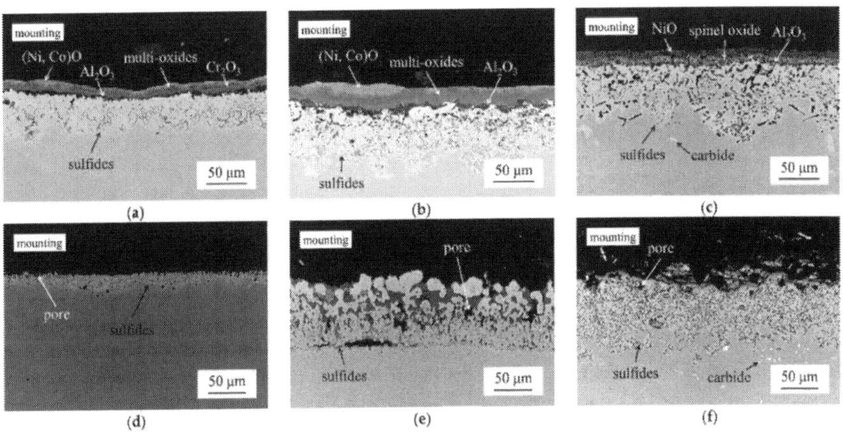

Figure 8. SEM-BSE images of samples after 900 °C/100 h (5 cycles) hot corrosion salt-coated tests, (a) HESA-1; (b) HESA-2; and (c) CM247LC; Cross section SEM-BSE images of 900 °C/20 h crucible tested samples; (d) HESA-1; (e) HESA-2; and (f) CM247LC.

In the salt-coated tests, the liquid film of molten salts on the top surface of specimens would gradually be depleted due to reaction with the substrate, followed by oxidation during the latter part of the 20 h exposure, so oxide scales would form above the corroded surfaces (Figure 8a–c). This kind of hot corrosion is discontinuous and coupled with oxidation, so the surfaces are less corroded. As for crucible tests, specimens were immersed into the salt solution, hence the salt supplementation can be abundant. The corrosion has been continuous during exposure and has led to more severe metal loss.

According to Figure 8, the sulfide penetration depth of HESA-1 can be less than that of HESA-2 and CM247LC in both tests, and the surfaces of HESA-1 are obviously less corroded, demonstrating its superior hot corrosion resistance.

4. DISCUSSION

The present study has shown that HESA-1 is a Cr_2O_3-former and HESA-2 can be an Al_2O_3-former, with respect to high temperature oxidation behavior. With high solute contents, both HESA alloys have exhibited formation of more complex types of oxides comparing to those of CM247LC. Figure 9 plots the schematic oxidation mechanism of HESA. Oxides scale of HESA-1 consists of outer (Ni, Co)O, $CoFe_2O_4$, $Fe_3Ti_3O_{10}$, Cr_2O_3, TiO_2 and Al_2O_3; while HESA-2 contains (Ni, Co)O, $CoFe_2O_4$, $(Ni, Ti)_3O_4$, $CrTi_2O_5$, $CrTaO_4$ and Al_2O_3. At 900 °C, since HESA-1 is not able to form continuous Al_2O_3, the thickness of the oxide scales has been increased from 8 μm (5 h) to 23 μm (100 h). By contrast, the oxidized region of HESA-2 has only increased slightly from 6 μm (5 h) to 12 μm (100 h) due to continuous Al_2O_3 protection, (Figure 2 and Figure 4). However, it appears that complex oxides such as $CoFe_2O_4$, $Fe_3Ti_3O_{10}$, $CoFe_2O_4$, $(Ni, Ti)_3O_4$, $CrTi_2O_5$, $CrTaO_4$ may have contributed to a more rapid increase in oxidation weight gain for HESA alloys (Figure 1). As the thickness of continuous Cr_2O_3 of HESA-1 and continuous Al_2O_3 of both HESA-2 and CM247LC increases, the rate of oxidation gradually decreases. At 1100 °C, the role of the protective Al_2O_3 has become more pronounced, as both Al_2O_3 formers HESA-2 and CM247LC have outperformed HESA-1 significantly, Figure 1b. Since HESA-2 can form a sufficient thickness of continuous Al_2O_3 rapidly, its overall oxide thickness has only increased slightly from 10 μm (5 h) to 16 (100 h) μm (Figure 2 and Figure 5). Therefore, outward diffusion has been hindered by the continuous Al_2O_3 layer and these outer complex oxides are not able to grow further. On the other hand, Cr2O3 is less effective against oxidation at higher temperatures. Since HESA-1 can only form discontinuous Al2O3, its oxidation weight gain has been increased dramatically at 1100 °C with time.

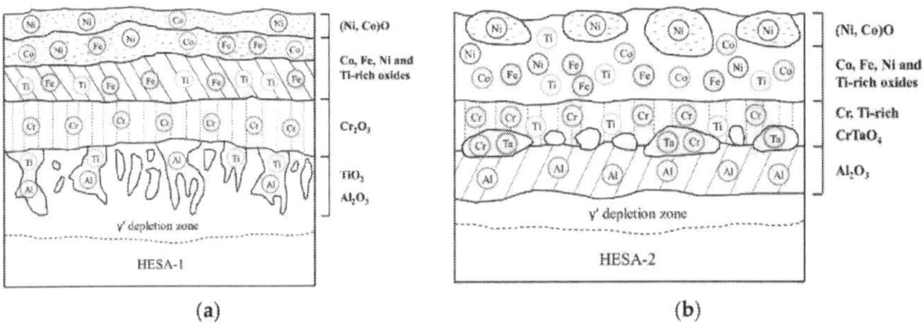

Figure 9. Illustrations of oxide formation on (a) HESA-1 and (b) HESA-2.

According to the selective oxidation mechanisms in the Ni-Cr-Al systems [34], HESA-1 can be categorized as a Type-II alloy, which forms mainly Cr2O3 with subscales of discontinuous Al2O3, and HESA-2 is a Type-III alloy, which forms mainly Al2O3 with no internal oxidation. The underlying mechanism of the selective oxidation is associated with the activities of the elements in the alloy system. The thermodynamic software PANDAT (Pan-Ni database version 8) [35] has been used in the present work to calculate the Al and Cr activities of HESA-1, HESA-2 and CM247LC at 900 °C and 1100 °C, Table 2. It is apparent that both HESA alloys possess higher Al activity than those of CM247LC, and this can be a result of the third-element effect. For example, higher addition of Cr in Ni-Cr-Al system can enhance Al activity to promote Al2O3 formation with lower content of Al [36]. Interestingly, although HESA-1 possesses the highest Al activity, which indicates the faster formation of Al_2O_3, this alloy still only forms a discontinuous Al_2O_3 layer. This is due to its low Al content (Table 1). Short-term oxidation tests were conducted on HESA-1 and HESA-2 at 1100 °C, and the cross-section oxide scales are shown in Figure 10. After 3 min exposure, HESA-1 has already formed the thin Al_2O_3, but there is no obvious sign of Al_2O_3 formation in HESA-2. The formation of Al_2O_3 in HESA-2 can be observed after 10 min exposure, and that in HESA-1 is clearly discontinuous. The 30 min test results show that gradual internal oxidation occurs in HESA-1, while the Al_2O_3 formation in HESA-2 becomes more continuous. According to the literature, the Al concentration in conventional Ni-based superalloys is usually maintained at 5–6 wt.% to promote continuous Al_2O_3 formation [10,37,38,39]. By contrast, HESA-1 only contains 3.9 wt.% of Al, and it does not form continuous Al_2O_3, though its Al activity is the highest among the three alloys. As for HESA-2, which contains similar levels of Al and Cr compared to that of CM247LC, its high content of Ti may also attribute to its high Al activity [22], although Ti-rich oxides may form rapidly during the initial stage of oxidation. Therefore, to further improve the oxidation resistance of high entropy superalloys, future alloy design will try to limit the content of Ti in order to minimize the formation of complex oxides, and further increase the Al activity by alloy design for rapid Al_2O_3 formation.

Regarding the hot corrosion behaviour, the discontinuous corrosion by a salt-coated test and continuous corrosion by a crucible test have been examined. The total weight loss from salt-coated and crucible tests after 900 °C/20 h exposure can be estimated by a number of experimental values, including the 20 h isothermal oxidation weight gain data of Figure 1 at 900 °C, weight loss after the first salt-coated cycle, the average dimension loss after crucible tests, sample size and density. During salt-coated tests, the top surface is mainly corroded while the four side surfaces are oxidized. The bottom surface is adjacent with the refractory brick, so no significant oxidation is expected. The total weight loss from the first salt-coated cycle, which excludes oxidation weight gain from four side surfaces is 8, 6 and 7 mg for HESA-1, HESA-2 and CM247LC, respectively. As for crucible tests, all six surfaces are corroded. The volume change and alloy density of specimens can be used to estimate the overall weight loss, and the values are 34, 445 and 460 mg for HESA-1, HESA-2 and CM247LC, respectively. The weight loss from crucible test is indeed much higher due to continuous corrosion attack. This agrees with the microstructure observations, as more severely corroded surfaces on HESA-2 and CM247LC are shown in Figure 8e,f.

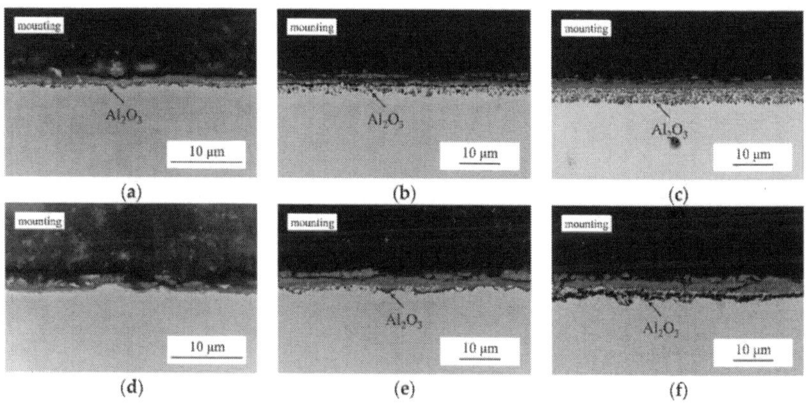

Figure 10. The oxidized HESA-1 at 1100 °C for (a) 3 min; (b) 10 min; and (c) 30 min; and that of HESA-2 at 1100 °C for (d) 3 min; (e) 10 min; and (f) 30 min.

Alloys with higher Cr content are known to perform better against hot corrosion, and the rate of sulphidation could also be significantly hindered [5,40]. The reason is attributed to Cr2O3 reacting to form several valence states such as NaCrO2, Na2CrO4 and Na2Cr2O7 with molten salts during corrosion. According to the sustained hot corrosion model, a negative solubility gradient between oxide/molten salts and salt/gas interfaces is required for corrosion to proceed [31,41]. The dissolution of Cr2O3 can result in a positive solubility gradient due to different oxygen activity at the interfaces [42], and this positive gradient would interrupt the hot corrosion mechanism. As for Al2O3, it does not exhibit multiple valence states, so the negative solubility gradient mechanism is sustainesd and leads to constant corrosion attack [31]. Therefore, HESA-1, which acts as a Cr2O3-forming alloy can resist hot corrosion more strongly than the Al2O3 former. In addition, it has reported that the diffusion velocity of sulfur in Co-bearing systems can be up to two orders of magnitude lower than that in Ni at high temperature [43,44], therefore the high Co content in HESA-1 may reduce the internal diffusion of sulfur and alleviate the subsequent hot corrosion degradation.

Table 2. The PANDAT calculated (Pan-Ni database) Al and Cr activity of HESA-1, HESA-2 and CM247LC at 900 and 1100 °C.

Activity		900 °C	1100 °C
HESA-1	Al ($\times 10^{-8}$)	5.88	20.75
	Cr ($\times 10^{-3}$)	5.18	2.24
HESA-2	Al ($\times 10^{-8}$)	4.32	19.46
	Cr ($\times 10^{-3}$)	3.72	1.64
CM247LC	Al ($\times 10^{-8}$)	0.75	6.69
	Cr ($\times 10^{-3}$)	3.52	1.67

With regard to the salt-coated hot corrosion test, after the first 20 h cycle, the weight losses of the present alloys are very similar (Figure 7a), and this is a result of the combined weight loss from corrosion on the top surface and weight gain from oxidation on the sides. According to the result of crucible test (Figure 7b), the weight loss of HESA-1 is much less than that of HESA-2 and CM247LC. Consequently, this indicates that the following oxidation weight gain on the top surface of HESA-2 and CM247LC after molten salt depletion should be more than that of HESA-1. Since HESA-2 and CM247LC are less protected against hot corrosion, the Al depletion beneath the sulfide front would be larger, and this leads to insufficient Al2O3 protection against the subsequent oxidation. According to Figure 8b,c, the external oxide thickness of HESA-2 and CM247LC are indeed greater than that of HESA-1, which agrees with the more severe post-corrosion oxidation. The weight loss becomes more significant from the second salt-coated cycle onward. During 20–40 h exposure, the re-coated salts would again attack the outer oxides which formed during the first cycle and further corrode the material within. The corrosion rate of HESA-1 can remain low with Cr2O3 protection, while the Al2O3 forming alloys HESA-2 and CM247LC exhibit more weight loss owing to increasing depletion of Al. With accumulated test cycles up to 100 h, much more severe depletion of Cr and Al near the top surface would occur, so the corrosion rates have been gradually enhanced due to lesser protection from Cr2O3 and Al2O3, resulting in an increase in weight loss during the later cycles, Figure 7a.

In this study, the high temperature surface stability of high entropy superalloys has been investigated. The increase in Al content from HESA-1 to HESA-2 can improve the oxidation resistance of HESA-2 to be comparable to that of CM247LC, while with high Cr content, HESA-1 can exhibit excellent resistance against hot corrosion. The potential of HESAs to offer great high temperature surface stability has been confirmed. For future alloy design of high entropy superalloys, the Al content of HESA-1 should be elevated for continuous Al2O3 formation, and Ti content can be kept lower or partly replaced by another γ'-forming element such as Nb.

5. CONCLUSIONS

The oxidation and corrosion behaviours at elevated temperatures of novel high entropy superalloys (HESAs) are studied. Although high content of various solutes in HESA leads to the formation of complex oxides, the high Cr and Al activities of HESAs can still promote the formation of protective chromia or alumina layers on the surface. The Cr2O3 former HESA-1 can exhibit excellent hot corrosion resistance, while the Al2O3 former HESA-2 possesses good resistance against high temperature oxidation. Therefore, the surface stability of HESAs in oxidizing and corrosive environments has been demonstrated.

ACKNOWLEDGMENTS

Authors would like to thank the financial support from Ministry of Science and Technology, Taiwan (R.O.C.), project grant number: 103-2221-E-214-035, 103-2218-E-007-019. Also authors would like to thank J.W. Yeh for discussion on the topic of high entropy related materials. T.K. Tsao would like to thank the NTHU-NIMS coorpetive graduate program for supporting his stay at NIMS.

AUTHOR CONTRIBUTIONS

All authors contributed extensively to this study. Te-Kang Tsao and An-Chou Yeh designed the experimental structure. Chen-Ming Kuo performed the directional solidification casting of present alloys. Hideyuki Murakami gave the technical support and conceptual advice. Te-Kang Tsao carried out the experiment. All authors commented on the manuscript at all stages.

REFERENCES

1. Furrer, D.; Fecht, H. Ni-based superalloys for turbine discs. JOM 1999, 51, 14–17.

2. Caron, P.; Khan, T. Evolution of Ni-based superalloys for single crystal gas turbine blade applications. Aerosp. Sci. Technol. 1999, 3, 513–523.

3. Yeh, A.C.; Tin, S. Effects of Ru and Re additions on the high temperature flow stresses of Ni-base single crystal superalloys. Scripta. Mater. 2005, 52, 519–524.

4. Pollock, T.M.; Tin, S. Nickel-based superalloys for advanced turbine engines: Chemistry, Microstructure, and Properties. J. Propul. Power 2006, 22, 361–374.

5. Reed, R.C. The Superalloys: Fundamentals and Applications; Cambridge University Press: Cambridge, UK, 2006.

6. MacKay, R.A.; Gabb, T.P.; Smialek, J.L.; Nathal, M.V. A New Approach of Designing Superalloys for Low Density. JOM 2010, 62, 48–54.

7. Koizumi, Y.; Kobayashi, T.; Yokokawa, T.; Zhang, J.X.; Osawa, M.; Harada, H.; Aoki, Y.; Arai, M. Development of next-generation Ni-base single crystal superalloys. Superalloys 2004.

8. Kawagishi, K.; Harada, H.; Sato, A.; Sato, A.; Kobayashi, T. The oxidation properties of fourth generation single-crystal nickel-based superalloys. JOM 2006, 58, 43–46.

9. Reed, R.C.; Tao, T.; Warnken, N. Alloys-By-Design: Application to Nickel-Based Single Crystal Superalloys. Acta Mater. 2009, 57, 5898–5913.

10. Sato, A.; Chiu, Y.L.; Reed, R.C. Oxidation of nickel-based single-crystal superalloys for industrial gas turbine applications. Acta Mater. 2011, 59, 225–240.

11. Evans, A.G.; He, M.Y.; Suzuki, A.; Gigliotti, M.; Hazel, B.; Pollock, T.M. A mechanism governing oxidation-assisted low-cycle fatigue of superalloys. Acta Mater. 2009, 57, 2969–2983.

12. Deb, D.; Iyer, S.R.; Radhakrishnan, V.M. A comparative study of oxidation and hot corrosion of a cast nickel base superalloy in different corrosive environments. Mater. Lett. 1996, 29, 19–23.

13. Tong, J.; Dalby, S.; Byrne, J.; Henderson, M.; Hardy, M. Creep, fatigue and oxidation in crack growth in advanced nickel base superalloys. Int. J. Fatigue 2001, 23, 897–902.

14. Eliaz, N.; Shemesh, G.; Latanision, R.M. Hot corrosion in gas turbine components. Eng. Fail. Anal. 2002, 9, 31–43.

15. Yeh, J.W.; Chen, S.K.; Lin, S.J.; Gan, J.Y.; Chin, T.S.; Shun, T.T.; Tsau, C.H.; Chang, S.Y. Nanostructured high-entropy alloys with multiple principal elements: Novel Alloy Design Concepts and Outcomes. Adv. Eng. Mater. 2004, 6, 299–303.

16. Hsu, C.Y.; Juan, C.C.; Wang, W.R.; Sheu, T.S.; Yeh, J.W.; Chen, S.K. On the superior hot hardness and softening resistance of AlCoCrxFeMo0.5Ni high-entropy alloys. Mater. Sci. Eng. A 2011, 528, 3581–3588.

17. Tsai, K.Y.; Tsai, M.H.; Yeh, J.W. Sluggish diffusion in Co-Cr-Fe-Mn-Ni high-entropy alloys. Acta Mater. 2013, 61, 4887–4897.

18. Tsai, M.H.; Yeh, J.W. High-entropy alloys: A Critical Review. Mater. Res. Lett. 2014, 2, 107–123.

19. Yeh, A.C.; Tsao, T.K.; Chang, Y.J.; Chang, K.C.; Yeh, J.W.; Chiou, M.S.; Jian, S.R.; Kuo, C.M.; Wang, W.R.; Murakami, H. Developing new

type of high temperature alloys—High Entropy Superalloys. Int. J. Metall. Mater. Eng. 2015.

20. Yeh, A.C.; Chang, Y.J.; Tsai, C.W.; Wang, Y.C.; Yeh, J.W.; Kuo, C.M. On the Solidification and Phase Stability of a Co-Cr-Fe-Ni-Ti High-Entropy Alloy. Metall. Mater. Trans. A 2014, 45A, 184–190.

21. Harris, K.; Erickson, G.L.; Schwer, R.E. MAR M 247 Derivations—CM 247 LC DS Alloy and CMSX Single Crystal Alloys: Properties & Performance. Superalloy 1984.

22. Yeh, A.C.; Yang, K.C.; Yeh, J.W.; Kuo, C.M. Developing an advanced Si-bearing DS Ni-base superalloy. J. Alloy. Compd. 2014, 585, 614–621.

23. Cryosystems, O. Crystallographica Search-Match. J. Appl. Crystallogr. 1999, 32, 379–380.

24. Das, D.; Singh, V.; Joshi, S. High temperature oxidation behaviour of directionally solidified nickel base superalloy CM–247LC. Mater. Sci. Technol. 2003, 19, 695–708.

25. Chiou, M.S.; Jian, S.R.; Yeh, A.C.; Kuo, C.M. Effects of Aluminum Addition on the High Temperature Oxidation Behavior of CM-247LC Ni-based Superalloy. Int. J. Electrochem. Sci. 2015, 10, 5981–5993.

26. Seal, S.; Kuiry, S.; Bracho, L.A. Studies on the surface chemistry of oxide films formed on IN-738LC superalloy at elevated temperatures in dry air. Oxid. Met. 2001, 56, 583–603.

27. Donachie, M.J.; Donachie, S.J. Superalloys: A Technical Guide; ASM International: Novelty, OH, USA, 2002.

28. Li, M.; Sun, X.; Jin, T.; Guan, H.; Hu, Z. Oxidation Behavior of a Single-Crystal Ni-Base Superalloy in Air—II: At 1000, 1100, and 1150 C. Oxid. Met. 2003, 60, 195–210.

29. Kumar, S.; Mudgal, D.; Singh, S.; Prakash, S. Cyclic oxidation behavior of bare and Cr3C2–25 (NiCr) coated super alloy at elevated temperature. Adv. Mater. Lett. 2013, 4, 754–761.

30. Liu, L.; Wu, S.; Dong, Y.; Lü, S. Effects of alloyed Mn on oxidation behaviour of a Co-Ni-Cr-Fe alloy between 1050 and 1250 °C. Corros. Sci. 2015, 104, 236–247.

31. Bose, S. High-Temperature Corrosion. In High Temperature Coatings; Bose, S., Ed.; Butterworth-Heinemann: Burlington, VT, USA, 2007; pp. 53–70.

32. Bornstein, N.S.; DeCrescente, M.A. The Role of Sodium and Sulfur in the Accelerated Oxidation Phenomena–Sulfidation. Corrosion 1970, 26, 309–314.

33. Pettit, F.; Meier, G.; Gell, M.; Kartovich, C.; Bricknel, R.; Kent, W.; Radovich, J. Oxidation and hot corrosion of superalloys. Metal. Soc. AIME Warrendale PA 1984, 651, 651–687.

34. Giggins, C.S.; Pettit, F.S. Oxidation of Ni-Cr-Al Alloys between 1000 Degrees and 1200 Degrees C. J. Electrochem. Soc. 1971, 118, 1782–1790.

35. Chen, S.L.; Daniel, S.; Zhang, F.; Chang, Y.A.; Yan, X.Y.; Xie, F.Y.; Schmid-Fetzer, R.; Oates, W.A. The PANDAT software package and its applications. Calphad 2002, 26, 175–188.

36. Stott, F.H.; Wood, G.C.; Stringer, J. The Influence of Alloying Elements on the Development and Maintenance of Protective Scales. Oxid. Met. 1995, 44, 113–145.

37. Birks, N.; Meier, G.H.; Pettit, F.S. Forming Continuous Alumina Scales to Protect Superalloys. JOM 1994, 46, 42–46.

38. Gordon, A.P.; Trexler, M.D.; Neu, R.W.; Sanders, T.J.; McDowell, D.L. Corrosion kinetics of a directionally solidified Ni-base superalloy. Acta Mater. 2007, 55, 3375–3385.

39. Suzuki, A.S.; Kawagishi, K.; Yokokawa, T.; Harada, H.; Kobayashi, T. A New Oxide Morphology Map: Initial Oxidation Behavior of Ni-Base Single-Crystal Superalloys. Metall. Mater. Trans. A 2012, 43, 155–162.

40. Mrowec, S.; Werber, T.; Zastawni, M. Mechanism of High Temperature Sulphur Corrosion of Nickel-Chromium Alloys. Corros. Sci. 1966, 6, 47–68.

41. Rapp, R.A.; Goto, K. The Hot Corrosion of Metals by Molten Salts. In Proceedings of the Second International Symposium on Molten Salts; Physical Electrochemistry Division, Electrochemical Society: Pennington, NJ, USA, 1981; p. 159.

42. Rapp, R.A. Chemistry and electrochemistry of hot corrosion of metals. Mater. Sci. Eng. 1987, 87, 319–327.

43. Beltran, A.M.; Shores, D.A. Hot Corrosion. In The Superalloys; Wiley-Interscience: New York, NY, USA, 1972; pp. 317–339.

44. Qiao, M.; Zhou, C.G. Hot corrosion behavior of Co modified NiAl coating on nickel base superalloys. Corros. Sci. 2012, 63, 239–245.

CHAPTER 5

Effects of Anodizing Parameters in Tartaric-Sulphuric Acid on Coating Thickness and Corrosion Resistance of Al 2024 T3 Alloy

Mohammad Zaki Mubarok[1*], Wahab[2], Sutarno[3], Soleh Wahyudi[4]

[1]Department of Metallurgical Engineering, Faculty of Mining and Petroleum Engineering, Institut Teknologi Bandung, Bandung, Indonesia

[2]Department of Chemistry, Faculty of Mathematic and Natural Sciences, Halu Oleo University, Kendari, Indonesia

[3]Mineral and Metallurgical Engineering Specialization, Postgraduate Program of Mining Engineering, Faculty of Mining and Petroleum Engineering, Institut Teknologi Bandung, Bandung, Indonesia

[4]PT. Rekayasa Plating, Cimahi, Jawa Barat, Indonesia

ABSTRACT

2024 T3 is one of aluminium alloys which are widely used in the aircraft structures. Anodizing of alluminium alloy in tartaric-sulphuric acid (TSA) electrolyte is developed to obtain more environmentally-friendly process and to produce anodize layer with better corrosion resistance. In this research work, the influences of anodizing parameters of Al 2024 T3 in TSA on the thickness, weight and corrosion resistance of the anodize layer are studied. Corrosion resistance test was carried out by conducting salt spray test for 336 hours and anodic polarization measurements using potentiostat.

Results of three-factor analysis of variance (ANOVA) demonstrated that the most influencing factor that determines the thickness and weight of the anodize layer is temperature, followed by applied voltage, duration of anodizing, voltage-temperature interaction, interaction of temperature-duration of anodizing, interaction of voltage-temperature-duration of anodizing, and interaction of voltage and duration of anodizing. The pit density and corrosion current density (i_{corr}) were found to be dependent on the coating thickness. The anodize layer with a thickness of higher than 3 μm was not experienced to pitting corrosion during 336 hours of salt spray test.

KEYWORDS

Aluminium Alloy, Anodizing, TSA, Corrosion, Polarisation, Pitting

1. INTRODUCTION

Aluminium alloy of 2xxx is widely used in aircraft structures due to its light weight and high specific strength [1] [2] . Alloying with elements such as copper improves mechanical properties of the alloy, but decreases its corrosion resistance due to microscopic galvanic effect [2] . In order to improve their corrosion resistance, anodizing of 2xxx series aluminium alloys in chromic acid has traditionally been used for application in aircraft structures. Chromic acid (CA) anodizing provides excellent corrosion protection to the alloy with relatively low impact on its fatigue life compared to anodization in other acid baths [2] [3] . Hexavalent chromium ions are the major component of CA electrolyte that is able to create conversion layers and gives corrosion protection in addition to that provided by aluminium oxide layer. Despite its excellent performance in providing the corrosion resistance of the anodize layer, hexavalent chromium species are known to be toxic and hazardous materials. Research efforts have been performed to replace chromates in surface treatment technology and to look for more environmentally-friendly anodizing processes [1] - [5] . During the last decade, numerous investigations of anodizing in sulphuric acid and boric acid have been made for replacing CA anodizing. Anodizing in dilute sulphuric acid (DSA) is performed to obtain thin anodize layer (1 - 5 mm) that gives lower impact on the fatigue strength of the alloy. However, despite an improvement of fatigue strength of the alloy, its corrosion resistance is still lower than the alloy being anodized in CA [6] .

Anodization of aluminium alloy in tartaric-sulphuric acid (TSA) is recent development for chromate-free anodizing that is regarded as environmentally-friendly, offering shorter duration of electrolysis and producing excellent corrosion resistance [1] - [3] . Following Curioni et al. [7] , tartaric acid ($C_4H_6O_6$) reduces the growth rate of porous oxide film during TSA anodizing without significantly affecting the mechanism of the porous film formation. It was reported that tartaric acid forms residue in pore solution of the anodize film which limits anodic reaction of aluminium dissolution in nearly neutral, chloride-rich environment [7] . Boisier et al. [6] have conducted FESEM and EIS study of sealed AA2024 T3 alloy anodized in sulphuric acid electrolyte to investigate the influence of tartaric acid on anodic film morphology and on corrosion resistance of hydrothermally sealed anodized alloy. It was concluded that the presence of tartaric acid results in the formation of anodic films with lower porosity than those produced by DSA anodizing. The EIS study revealed that corrosion resistance of sealed anodize film in TSA is significantly better than the sealed film obtained by DSA anodizing which is mainly associated with the higher compactness of the porous layer and higher resistance of the hydrothermally- sealed barrier layer [6] .

Despite known beneficial effect of tartaric acid addition in sulphuric acid bath in improving corrosion resistance of the aluminium alloy, the effect of anodizing parameters during TSA anodizing (i.e. temperature, voltage, and duration of electrolysis) on the coating properties is still poorly reported. At the present paper, the effects of anodizing parameters of Al 2024 T3 Alloy in TSA solution on the coating weight, thickness, surface morphology and its corrosion resistance are discussed. Three-factor analysis of variance method (ANOVA) is conducted to evaluate the most influential parameters that affect the coating thickness and weight as well as its corrosion resistance.

2. EXPERIMENTAL

The specimens, chemicals for surface preparation and the electrolyte were kindly donated by PT. Dirgantara Indonesia, the only aircraft manufacturing company in Indonesia. Chemical composition of AA 2024 T3 aluminium alloy used in the investigations is presented in Table 1. The specimens have dimensions of 80 × 30 × 1 mm^3. Prior to be anodized, the specimens were subjected to series of surface preparation. The specimens were polished by SiC abrasive paper (i.e. from 1000 to 2000 grid), ultrasonic-cleaned in alcohol 95%, washed with distilled water and then air dried. Degreasing was then carried out by using a commercial alkaline cleaner bath of Turco 4215 NCLT® at 50°C for 15 minute. At the final stage of the surface preparation, the specimens were dipped in Ardrox 295 GD® at room temperature for 5 minute. The major component of Ardrox 295 GD® solution is nitric acid (HNO_3) which was used to remove residual oxide layer on the surface of the specimens prior to be anodized. After each step of the surface preparation, the specimens were rinsed by distilled water.

Series of anodizing experiments in TSA solution were carried out under variations of applied voltages (10 V, 15 V, 20 V), temperatures (20°C, 30°C, 40°C) and duration of anodizing (20, 40, 60 minute). The electrolyte used was 1500 mL TSA which contains 44.7% (v/v) tartaric acid ($C_4H_6O_6$) 80 wt% and 21.7% (v/v) 40 wt% sulphuric acid (H_2SO_4). After anodizing was finished, sealing was performed by immersing the anodized specimen in Alodine® solution for 1 minute. The thickness of anodize layer was measured at three different points, namely on the top, middle and bottom of the surface of the anodized specimens by using elcometer that works based on eddy current measurements. Analysis using scanning electron microscope (SEM) and energy-disper- sive X-ray (EDX) was performed to elucidate the surface morphology of the anodize layer and its chemical composition.

Table 1. Chemical composition of AA2024 T3 aluminium alloy used in the investigation.

Component	(wt%)
Cu	4.45
Mg	1.57
Mn	0.56
Fe	0.17
Zn	0.16
Si	0.06
Others	0.04
Al	Balance

Corrosion resistance of the anodize layer was measured by conducting salt spray tests for 336 hour and cyclic- anodic polarization measurements using potentiostat. The salt spray test was aimed at evaluating the resistance of the anodize layer to pitting corrosion and measuring the pit density (i.e. the number of pit per unit area), while polarization measurements using potentiostat is done to obtain corrosion parameter data which include corrosion current density (i_{corr}), corrosion potential (E_{corr}) and pitting potential (E_{pit}). The specimens for polarization measurements were mounted in resin and have active area of 1 cm^2 prior to be anodized and used in the polarization measurements. The polarization measurements were done in solution which contains 35.6 gram/liter of NaCl. saturated calomel electrode (SCE) and a graphite rod were used as reference electrode and counter-electrode, respectively. Three-factor analysis of variance (ANOVA) was performed to evaluate the significance influences of applied voltage (A), temperature (B) and duration of anodizing (C), as well as the interactions between these variables on the thickness, weight and corrosion resistance of the anodize layer.

3. RESULTS AND DISCUSSION

3.1. Correlation between Anodizing Current Density and Applied Voltage

Applied voltages were raised from open circuit potential (OCP) with a sweep potential of 2 V/min up to holding voltages of 10, 15, and 20 V for 20, 40, and 60 minute at 20°C, 30°C, and 40°C. Current flow through the cell during anodizing was measured. Profiles of current density versus time recorded during the first 300 minute of anodizing under variations of applied voltages at 20°C and averages current density versus applied voltages at temperatures of 20°C, 30°C and 40°C are depicted in Figure 1(a) and Figure 1(b), respectively. Profiles of current density follows general trend of that commonly observed during anodizing of aluminium alloy. The alterations of current density during initial periods of anodizing indicate stages occur during the formation of anodize layer. As applied voltage was increased from 0 to target voltages (i.e. 10, 15 and 15 V) with ram rate of 2 V/min, the currents fluctuate. This fluctuation indicates simultaneous formations of barrier layer and porous layer on the surface of the specimen. The porous layer forms due to partial dissolution of Al_2O_3 layer in acidic solution. As the voltage reaches the targeted voltage, the current density slightly increases then decreases and finally constant at certain level as illustrated by Figure 1(a). At a constant current density, the growth of barrier layer and porous layer performs at steady state. The current density increases by the increased of applied voltage and anodizing temperature as illustrated by Figure 1(b). Increasing temperature reduces electrolyte resistance that consequently enhances current flow through the circuit at certain voltage.

3.2. Effect of TSA Anodizing Parameters on Thickness and Weight of the Anodize Layer

Applied voltage is one of key factors that determine the thickness and weight of anodize layer. The influences of applied voltage on the thickness and weight of anodize layer are shown in Figure 2 and Figure 3, respectively. The increase of applied voltage resulted in an enhanced of current density as has been illustrated in Figure 1(b).

Current flows through electrolyte are associated with ionic mobility in the electrolyte. The formation of anodize layer (Al_2O_3) is contributed by O^{2-} ion from water decomposition ($3H_2O \rightarrow 3O^{2-} + 6H^+$) in the interface of anodize layer and the electrolyte and Al^{3+} from the oxidation of the surface of aluminium alloy ($2Al \rightarrow 2Al^{3+} + 6e^-$) takes place at the interface of the alloy and the oxide layer. A higher applied voltage leads to a higher O^{2-} and Al^{3+} mobility and resulted in the higher coating thickness and weight after certain time of anodizing as can be seen in Figure 2 and Figure 3.

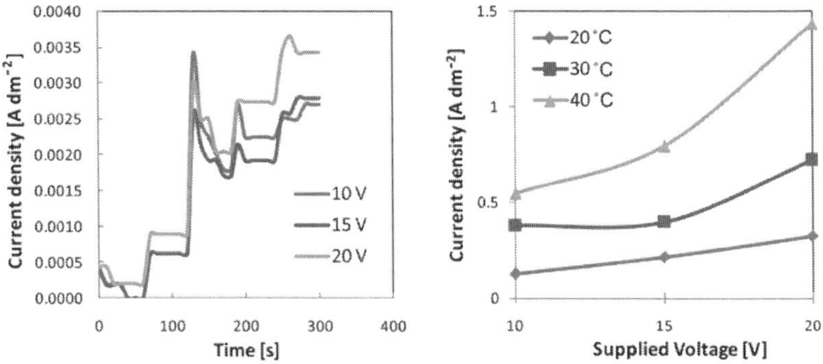

Figure 1. (a) Profiles of current density versus time recorded during the first 300 minute of anodizing under variations of applied voltages at 20°C; (b) Averages current density versus applied voltages at temperatures of 20°C, 30°C and 40°C.

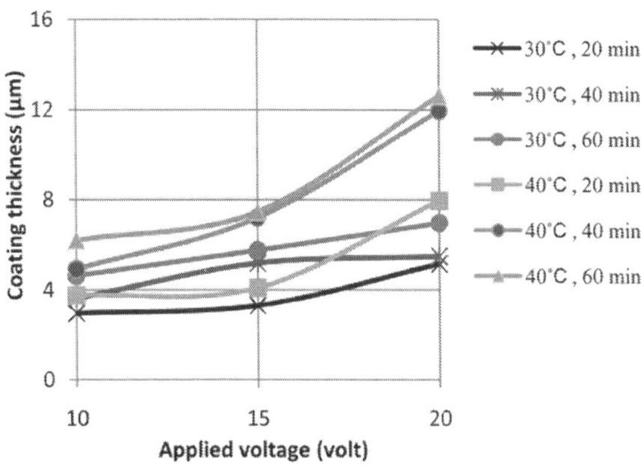

Figure 2. Profiles of coating thickness as a function of applied voltage at temperatures of 30°C and 40°C and various anodizing time.

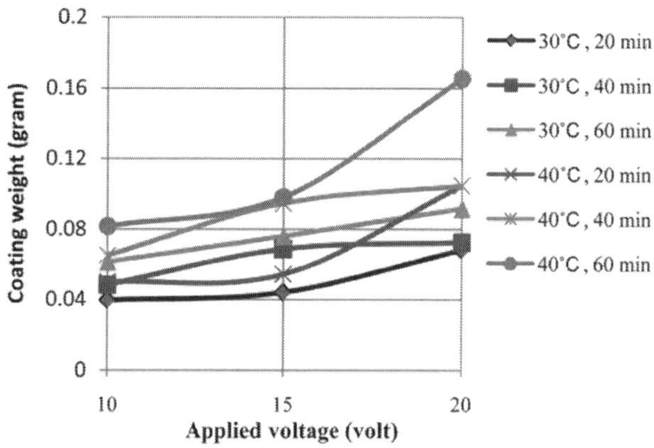

Figure 3. Profiles of coating weight as a function of applied voltage at temperatures of 30°C and 40°C and various anodizing time.

Figure 4 and Figure 5 show the effect of temperature on the thickness and weight of anodize layer, respectively. It can be seen that the increase of temperature tends to increases the thickness and weight of anodize layer. The effect of anodizing temperature is associated with the mass transfer rates of O^{2-} and Al^{3+} ions during coating formation. The increased of temperature accelerates inward diffusion of O^{2-} ions through oxide layer that further reacts with Al^{3+} ions to form Al_2O_3 layer at the interface of the alloy and the oxide layer. Similarly, the enhanced of temperatures also accelerates outward diffusion of Al^{3+} ions from the surface of the alloy to the interface of oxide layer-electrolyte that further react with O^{2-} ions at the interface of oxide layer-electrolyte. However, there has been a certain level of anodizing temperature, beyond which the increased of temperature would give negative impact to the thickness and weight of anodize layer due to overheating. At this condition, the dissolution of anodize layer takes preferentially than the formation of the anodize layer, thus results in porous anodize layer with lower thickness [8] [9] .

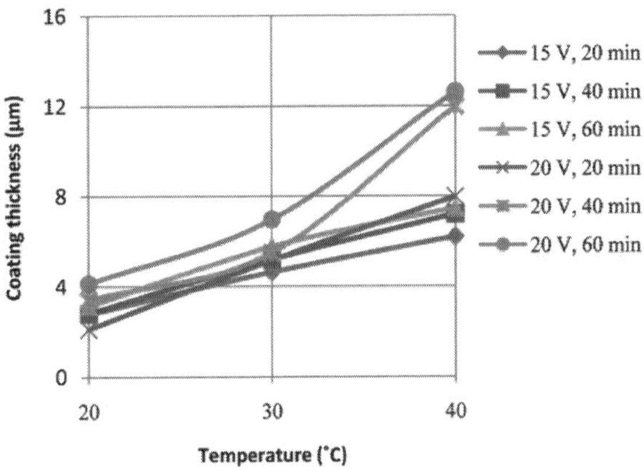

Figure 4. Profiles of coating thickness as a function of anodizing temperatures and duration of anodizing at applied voltage of 15 and 20 V.

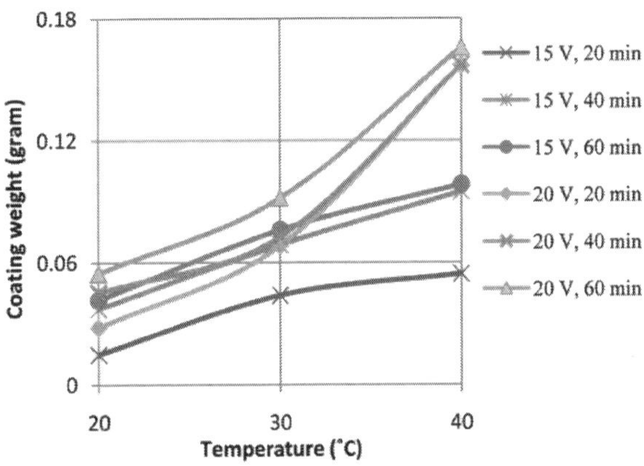

Figure 5. Profiles of coating weight as a function of anodizing temperature and duration of anodizing at applied voltages of 15 and 20 V.

Based on the results of three-factor analysis of variance (ANOVA), the most influential factor that determines the thickness and weight of the anodize layer is temperature (B), followed by applied voltage (A), duration of anodizing (C), interaction of voltage and temperature (AB), interaction of temperature and anodizing duration (BC), interaction of voltage-temperature-anodizing duration (ABC), and interaction of voltage and anodizing duration (AC). Correlations between applied voltage (x_1), duration of anodizing (x_2) and interaction between these parameters (x_1x_2) and the thickness and weight of anodize layer for anodizing temperatures of 20°C, 30°C and 40°C are presented in Table 2 and Table 3, respectively.

3.3. Effect of TSA Anodizing Parameters on Corrosion Resistance of the Anodize Layer

Applied voltage during TSA anodizing was found to have notable influence on the corrosion resistance of Al 2024 T3 aluminium alloy. Cyclic-anodic polarization curves of un-anodized and anodized specimens at various applied voltage in 35.6 g/l NaCl solution at room temperatures are presented in Figure 6. It can be seen that anodized specimens have significantly lower corrosion current density (i_{corr}) in comparison to that of un-ano- dized. The specimen being anodized at applied voltage of 10 V, temperature 30°C for 60 minute exhibits the lowest corrosion current density among the anodized specimens with various voltages at similar anodizing temperature and time. Polarization behaviour of the anodized specimens is correlated with the quality of the anodize layer, specifically its porosity. As the anodize layer of Al_2O_3 is electrically non-conductive, more porous layer will result in higher corrosion and passivation current densities. In this regard, anodizing at 10 V, temperature 30°C for 60 minute gives the best anodize layer that have the lowest corrosion and passivation current densities in comparison to those done at 15 and 20 V under similar temperature and anodizing time. Following the investigation results of Sabry and Alobaid [9] , anodizing at lower voltage leads to the formation of thicker barrier layer and smaller pore size. Smaller pore size reduces the sites that can be occupied by chloride ions that initiate pitting corrosion. In addition to the anodizing variables, the effectiveness of sealing after anodizing would also affect the corrosion resistance of the anodize layer. If the sealing process does not properly close the pores of anodize layer, chloride ions will penetrate and being accumulated within the pores of the anodize layer and starts to depassivate the anodize layer.

Figure 7 shows cyclic-anodic polarization curves of un-anodized and anodized specimens from anodizing at various temperatures, applied voltage of 10 V and anodizing duration of 40 minute. It was found that anodizing temperature of 30°C gives the best corrosion resistance of the anodized specimen which is represented by the lowest corrosion current density, followed by anodizing at 20°C and 40°C. Cyclic-anodic polarization measurements were also carried out on anodized specimens resulted from the anodizing with various durations at applied voltage of 15 V and temperature 40°C. It was found that thicker anodize layers resulted by longer anodizing times exhibits better corrosion resistance indicated by lower corrosion current densities. It is clearly understood that the thicker the anodize layer, the longer the aggressive ions reach the surface of the specimen. However, the thickness of the anodize layer is not the solely factor that determines the corrosion resistance of the specimen and it is still dependent on the quality of the anodize layer, specifically its porosity. Based on the thickness data in which R is correlation coefficient and S represents error level of the calculation obtained at various applied voltage, temperature and anodizing time, the thickness of anodize layer has low correlation with corrosion current density with R^2 of 0.057 as can be seen in Figure 8.

Table 2. Correlation between applied voltage, duration of anodizing and interaction between them with the thickness of anodize layer for anodizing temperatures of 20°C, 30°C and 40°C.

Temperature (°C)	Regression Equation	R^2	S
20	$y = -1.164 - 0.178x_1 + 0.1697x_2 + 0.00933\,x_1^2 - 0.001542\,x_2^2 + 0.000375x_1x_2$	0.9835	0.1621
30	$y = -0.42 + 0.214x_1 + 0.0460x_2 + 0.00011\,x_1^2 + 0.000111\,x_2^2 - 0.00013x_1x_2$	0.9435	0.3678
40	$y = 10.88 - 1.845x_1 + 0.1862x_2 + 0.07178\,x_1^2 - 0.002431\,x_2^2 + 0.0064x_1x_2$	0.9864	0.4208

Table 3. Correlation between applied voltage, duration of anodizing and interaction between them with the weight of anodize layer for anodizing temperatures of 20°C, 30°C and 40°C.

Temperature (°C)	Regression Equation	R^2	S
20	$y = -0.0137 - 0.00231x_1 + 0.002206x_2 + 0.000121\,x_1^2 - 0.000020\,x_2^2 + 0.000005x_1x_2$	0.9818	0.0022
30	$y = -0.0041 + 0.00281x_1 + 0.00006x_2 + 0.000001\,x_1^2 + 0.000001\,x_2^2 - 0.000002x_1x_2$	0.9425	0.0048
40	$y = 0.1433 - 0.02402x_1 + 0.002418x_2 + 0.000934\,x_1^2 - 0.000032\,x_2^2 + 0.000084x_1x_2$	0.9864	0.0054

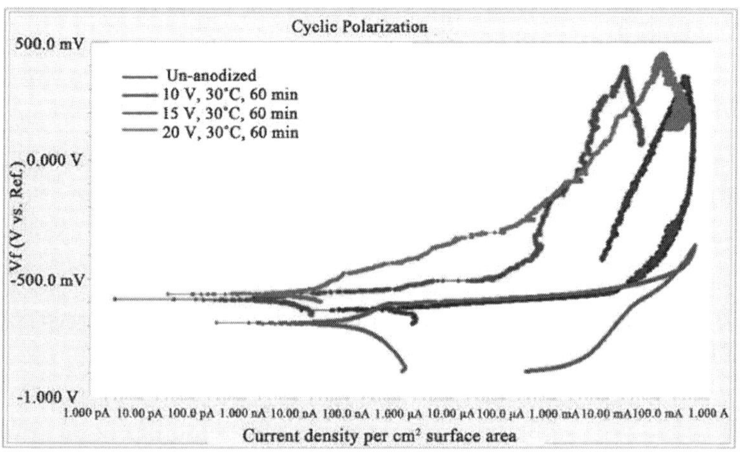

Figure 6. Cyclic-polarization of un-anodized and anodized specimens from anodizing at various applied voltages at 30°C for 60 minute.

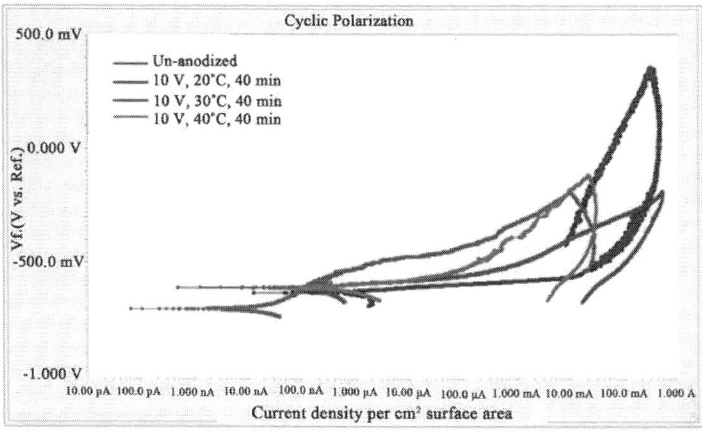

Figure 7. Cyclic polarization of un-anodized and anodized specimens under various temperatures at applied voltages of 10 V and anodizing time 40 minute.

Results of salt spray tests for 336 hour show that the anodized specimens with thicknesses of more than 3 μm did not exhibit pitting (Figure 9). This is in contrast to un-anodized specimen which was severely experienced to chloride attack. Based on ANNOVA result, it was found that anodizing temperature is the most decisive factor determined the resistance of the anodize layer to pitting corrosion. The highest pit density was obtained on the anodize layer from the anodizing at temperature of 20°C, while the anodize layers produced by anodizing temperatures of 30°C and 40°C were not experienced to pitting corrosion after 336 hour of salt spray tests. Based on ANNOVA result, correlation between applied voltage (x_1) and duration of anodizing (x_2) as well as interaction between them and the pit density (y) for anodizing temperature of 20°C is as follow:

$$y = 1.192 - 0.1236x_1 - 0.00469x_2 + 0.003056x_1^2 - 0.000043x_2^2 + 0.000417x_1x_2$$

The salt spray test results demonstrated that TSA anodizing of Al 2024 T3 specimens resulted in the anodize layer which meets the standard specification of Military Specification Anodizing (MIL-A-8625) with pit number of less than 15 per 9 dm^2 or 5 pit per 2 dm^2 after being subjected to salt spray test for 336 hour.

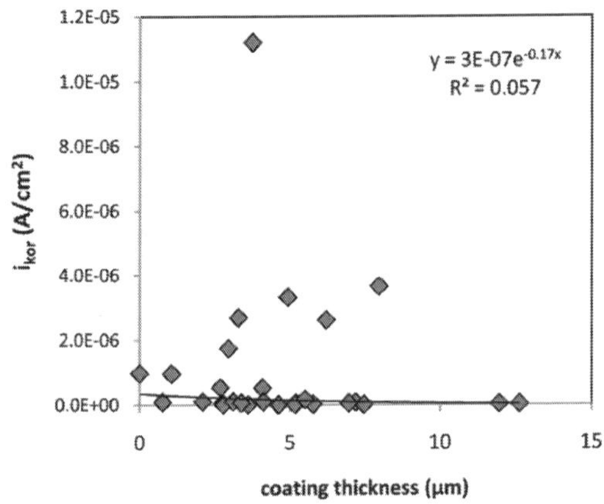

Figure 8. Correlation between corrosion current density versus coating thickness.

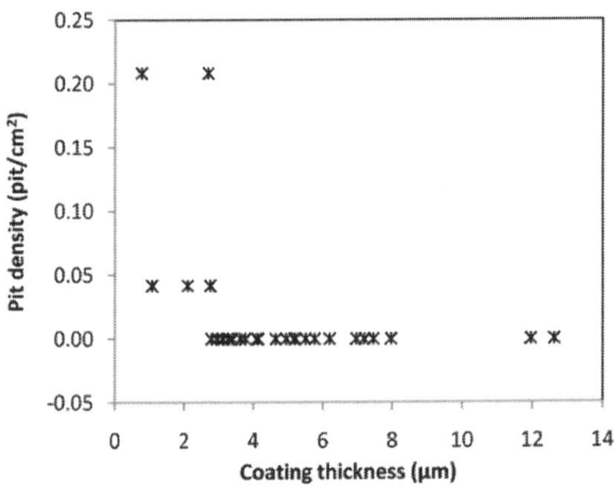

Figure 9. Correlation between pit density versus coating thickness.

3.4. Surface Morphology of the Anodize Layer

Effect anodizing temperature on the surface morphology of the anodize layer was evaluated by surface analysis of the anodized specimens from different temperature by scanning electron microscope (SEM). SEM micrographs of the surface of anodized layers obtained from anodizing at various temperatures, applied voltage of 15 volt for 40 minutes are presented in Figures 10(a)-(c). The surface of anodized specimen from anodizing at 20°C exhibits the most compact layer with fewer surface pores. At lower anodizing temperature, dissolution of oxide layer takes place slower than that at higher temperatures which results in fewer pores and more compact layer. Microcracks were detected on the surface of anodize layer from anodizing at 40°C (Figure 10(c)). These microcracks are associated with local heat effect on the surface of the anodize layer at high local current density and temperature. The local heat causes an inhomogeneous electric field distribution that leads to local breakdown of the oxide layer [10] [11] . Hence, it is demonstrated that temperature is the most decisive factor that must be carefully controlled during anodizing.

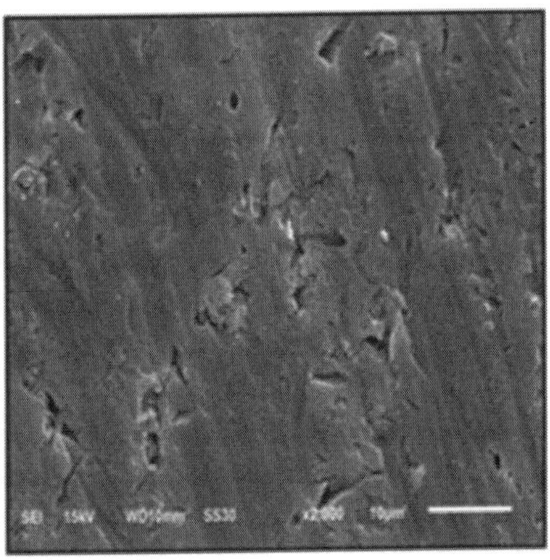

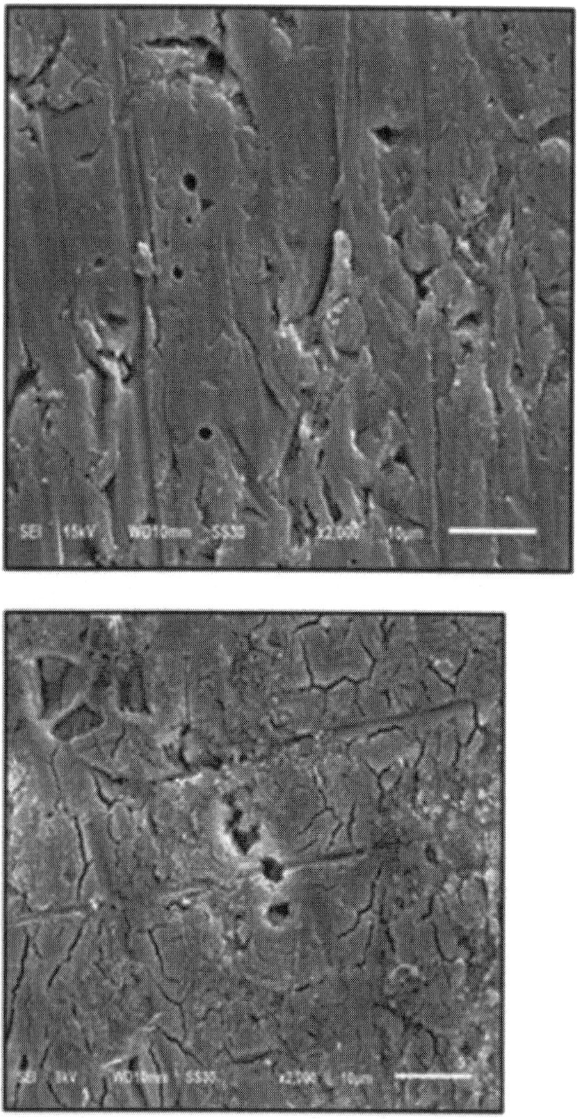

Figure 10. SEM micrographs of the surface of anodize layer obtained from anodizing with applied voltage of 15 volt for 40 minutes at (a) 20°C; (b) 30°C and (c) 40°C.

4. CONCLUSION

Results of three-factor analysis of variance (ANOVA) revealed that the most influencing factor determining the thickness and weight of the anodize layer is temperature, followed by applied voltage, duration of anodizing, interaction of voltage and temperature, interaction of temperature and anodizing duration, interaction of voltage- temperature and anodizing duration and interaction of applied voltage-anodizing duration. Corrosion current density (i_{corr}) and pit density were found to be dependent on the coating thickness. The anodize layer with a thickness of higher than 3 μm was not experienced to pitting during 336 hours of salt spray test. The anodizing process of Al 2024 T3 aluminium alloy in TSA offers an alternative to chromic acid which uses organic acid rather than poisonous chromium species, operated at lower temperatures and results in the anodize layer which has excellent corrosion resistance. Based on the data of pit density obtained from salt spray test for 336 hours, the anodize layer meets the standard specification of MIL-A-8625 with pit number of less than 15 per 9 dm^2 or 5 pit per 2 dm^2.

ACKNOWLEDGEMENTS

The authors thank PT. Dirgantara Indonesia for providing Al 2024 T3 specimen, chemicals for surface treat- ments and the electrolyte used in this investigation.

REFERENCES

1. Rubio, M.G., et al. (2009) Influence of Molybdate Species on the Tartaric Acid/Sulphuric Acid Anodic Films Grown on AA2024 T3 Aerospace Alloy. Corrosion Science, 51, 2034-2042.

2. Capelossi, V.R., Poelman, M., Hernandez, R.P.B., Melo, H.G.D. and Olivier, M.G. (2014) Corrosion Protection of Clad 2024 Aluminum Alloy Anodized in Tartaric-Sulfuric Acid Bath and Protected with Hybrid Sol-Gel Coating. Electrochimica Acta, 124, 60-79.

3. Rubio, M.G., et al. (2009) Effect of Postreatment on the Corrosion Behaviour of Tartaric-Sulphuric Anodic Films. Electrochimica Acta, 54, 4789-4800.

4. Harscoet, E. and Froelich, D. (2008) Use of LCA to Evaluate the Environmental Benefits of Substituting Chromic Acid Anodizing (CAA). Journal of Cleaner Production, 16, 1294-1305.

5. Zhang, J., Zhao, X., Zuo, Y. and Xiong, J. (2008) The Bonding Strength and Corrosion Resistance of Aluminum Alloy by Anodizing Treatment in a Phosphoric Acid Modified Boric Acid/Sulfuric Acid Bath. Surface & Coating Technology, 202, 3149-3156.

6. Boisier, G., Pébère, N., Druez, C., Villatte, M. and Suel, S. (2008) FESEM and EIS Study of Sealed AA2024 T3 Anodized in Sulfuric Acid Electrolytes: Influence of Tartaric Acid. Journal of the Electrochemical Society (JES), 155, 521-529.

7. Curioni, M., Skeldon, P., Koroleva, E., Thompson, G.E. and Ferguson, J. (2009) Role of Tartaric Acid on the Anodizing and Corrosion Behavior of AA 2024 T3 Aluminum Alloy. Journal of the Electrochemical Society (JES), 156, 147-153.

8. Theohari, S. and Kontogeorgou, C. (2013) Effect of Temperature on the Anodizing Process of Aluminum Alloy AA 5052. Applied Surface Science, 284, 611-618.

9. Sabry, R.S. and Alobaid, R. (2012) Effect of Preparation Conditions on Anodic Aluminium Oxide (AAO) Shape's and Size. International Journal of Science and Research (IJSR), 3, 866-872.

10. Aerts, T., Jorcin, J.B., Graeve, D.I. and Terryn, H. (2010) Comparison between the Influence of Applied Electrode and Electrolyte Temperatures on Porous Anodizing of Aluminium. Electrochimica Acta, 55, 3957-3965.

11. Aerts, T., De Graeve, I. and Terryn, H. (2008) Study of Initiation and Development of Local Burning Phenomena during Anodizing of Aluminium under Controlled Convection. Electrochimica Acta, 54, 270-279.

CHAPTER 6

Predicted Fracture Behavior of Shaft Steels with Improved Corrosion Resistance

Goran Vukelic [1,*] and Josip Brnic [2]

[1] Department of Marine Engineering and Ship Power Systems, Faculty of Maritime Studies Rijeka, University of Rijeka, Rijeka 51000, Croatia

[2] Department of Engineering Mechanics, Faculty of Engineering, University of Rijeka, Rijeka 51000, Croatia

ABSTRACT

One of the crucial steps in the shaft design process is the optimal selection of the material. Two types of shaft steels with improved corrosion resistances, 1.4305 and 1.7225, were investigated experimentally and numerically in this paper in order to determine some of the material characteristics important for material selection in the engineering design process. Ultimate tensile strength and yield strength have been experimentally obtained, proving that steel 1.4305 has higher values of both. In addition, J-integral is numerically determined as a measure of crack driving force for finite element models of standardized fracture specimens (single-edge notched bend and disc compact tension). Obtained J values are plotted versus specimen crack growth size (Δa) for different specimen geometries (a/W). Higher resulting values of J-integral for steel 1.4305 as opposed to 1.7225 can be noted. Results can be useful as a fracture parameter in fracture toughness assessment, although this procedure differs from experimental analysis.

KEYWORDS

steel 1.4305; steel 1.7225; fracture

1. INTRODUCTION

Material selection is a crucial step in the process of engineering design. Optimal selection of the material can significantly reduce the possibility of failures, along with understanding the nature and stress intensity that occurs in a designed structure. Engineering practices usually distinguishes one or few causes of failure: excessive force and/or temperature-induced elastic deformation, yielding, fatigue, corrosion, creep, etc. Selection of improper materials may have a negative effect on operational lifetime cycle and result in flaw appearance, which can cause structural failure.

A successful material-selection process implies reconciling requirements, such as appropriate strength of a material, sufficient level of rigidity, heat resistance, etc. For structures susceptible to crack growth, it is necessary to ensure that the material has been selected on the basis of fracture mechanics parameters.

Considering shaft design, the fracture mechanics approach must be used in order to account for high stresses and harsh operating conditions. Implementation of the fracture mechanics approach has the benefit of reducing potential failures, such as the fatigue induced fracture presented in a study of marine main engine crankshaft failure [1]. The agitator steel shaft failed due to an inadequate design, which was incapable of withstanding torsional-bending fatigue during operation [2]. The gearbox shaft failure occurred due to high stress concentrations at the corners of the wobbler of the shaft, causing fatigue crack initiation [3]. Improved design and machining practice suggested that this would help to prolong service life of the component. A forklift collapsed due to failure of axle shaft, caused by material inclusions and poor heat treatment [4].

Most of the mentioned failures occurred on steels typically used in the manufacturing of shafts intended for use in harsh environments, where a higher corrosion resistance is necessary. To be able to properly choose a suitable material for such an environment, characterization of a material is essential.

Fracture mechanics parameters that define material resistance to crack propagation are usually determined through experimental investigations of the material under consideration. Fracture behavior is usually estimated using some of the well-established fracture parameters, such as stress intensity factor (K), J-integral, or crack tip opening displacement (CTOD). J-integral is appropriate for quantifying material resistance to crack extension when dealing with ductile fracture of metallic materials, which includes nucleation, growth, and coalescence of voids [5]. For a growing crack, J-integral values can be determined for a range of crack extensions (Δa) and can be presented in the form of the J-resistance curve. This curve is usually obtained experimentally following standardized procedures, but it can be successfully complemented or even substituted by numerical methods, e.g., the finite element (FE) method. Some of the recent articles on this topic include discussion on the accuracy of J-integral obtained by experiments, two-dimensional (2D) FE analysis, three-dimensional (3D) FE analysis, or the Electric Power Research Institute method [6]. J-integral and CTOD are related through plastic constraint factors evaluated using 3D FE analyses of a clamped, single-edge tension specimen [7]. Methodology to evaluate 3D J-integral for finite strain elastic-plastic solid using FE analysis is proposed [8]. Stress intensity factors and T-stress of 3D interface cracks and notches are computed using the scaled boundary FE method [9].

This paper presents a comparison of numerically predicted J-values taken from the measure of crack driving force for two types of steel commonly used in shaft manufacturing, steels 1.4305 and 1.7225. Obtained material data may help designers to find the best solution in appropriate material selection.

2. MATERIALS AND METHODS

2.1. Considered Materials

The two materials compared are stainless steel 1.4305 (AISI 303) and alloy steel 1.7225 (AISI 4140). Steel 1.4305, commonly named chromium-nickel steel, is a derivative of a common grade stainless steel 1.4301, but with improved machinability. Chromium-molybdenum steel 1.7225 has an excellent strength to weight ratio, is readily machinable, and suitable for forging between 900 and 1200 °C.

The two materials differ substantially with respect to chemical composition (Table 1).

Table 1. Chemical composition of considered materials (wt%).

Material	C	Cr	Mn	Si	Mo	S	P
1.4305	0.047	17.4	2.0	0.584	–	0.252	0.0323
1.7225	0.45	1.06	0.74	0.32	0.17	0.018	0.014
Material	**Ni**	**Al**	**V**	**Nb**	**W**	**Cu**	**Rest**
1.4305	7.95	–	–	–	–	–	71.734
1.7225	0.04	0.02	0.01	0.02	0.02	0.04	97.078

Comparing the composition of steel 1.4305 to standard EN 10088-2:2005, it can be noted that the nickel percentage in the considered steel is just below the standard range (8%–10%), while manganese equals the maximum standard value (2%). As for steel 1.7225, comparing it to standard EN 10083-3:2006, it can be noted that carbon is equal to the maximum standard value (0.45%), while other elements are in standard ranges.

Both steels are widely considered for shafts intended to be used in marine, the petro or chemical industries, or as vehicular components, because both offer improved corrosion resistance, but with 1.4305 having significantly better resistance in a corrosive environment due to its elevated chromium content. Considering the differences in composition and corrosion resistance, it can be concluded that the two materials correspond to somewhat different ranges of specific applications. Steel 1.7225 can be found in bridge crane shafts [10], which are prone to fatigue failure, marine diesel engine crankshafts [11], where a material has to be adequate to severe working conditions, automotive applications [12], where axle shafts are sensitive to improper heat treatment, or in diesel engines of commercial vehicles [13], in which crankshafts need to be machined properly in order to avoid fatigue fractures. Chromium-nickel stainless steels can be found in agitator steel shafts [2] or mixer unit shafts [14], which prone to intergranular stress cracking at weld heat affected zones.

The equipment used to determine the mechanical properties of the materials was a computer-directed, materials testing machine (Zwick/Roell 400 kN, Zwick/Roell, Ulm, Germany) (Figure 1). Specimens were manufactured from rods made of the considered steels. Specimen geometry and uniaxial tensile test procedure were set, according to appropriate American Society for Testing and Materials (ASTM) standard [15]. Experimentally-obtained stress-strain curves are shown in Figure 2, and values of yield strength (σYS) and tensile strength (σTS) of considered materials are given in Table 1.

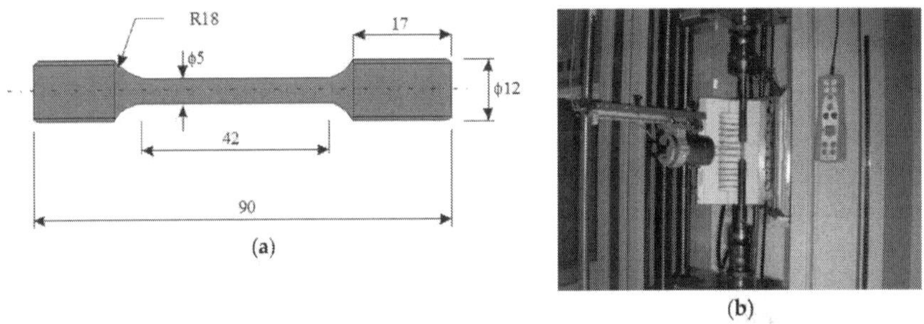

Figure 1. (a) Test specimen (dimension in mm); (b) testing machine.

When dealing with the fracture resistant design of structures, fracture toughness has an importance similar to yield strength when dealing with structure design against plastic deformation. The simple Charpy procedure for impact energy determination can be used to determine the fracture parameters. On the basis of known Charpy V-notch (CVN), impact energy correlation with fracture toughness can be made, e.g., with Roberts-Newton formula, independent of CVN energy range and temperature level [18]:

$$K_{Ic} = 8.47(CVN)^{0.63} \tag{1}$$

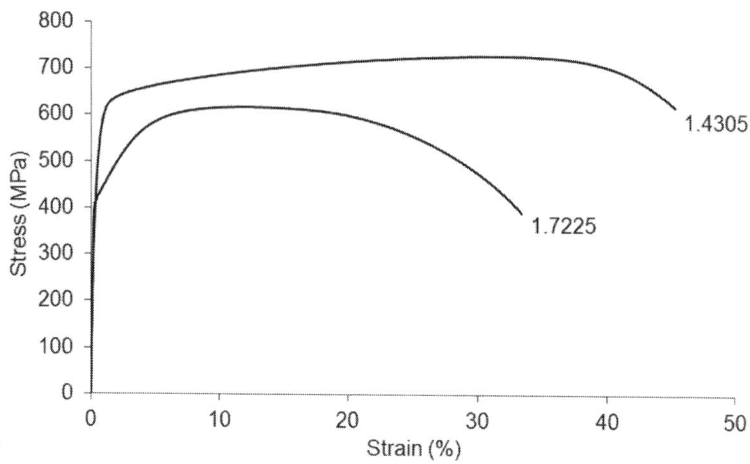

Figure 2. Uniaxial engineering stress-strain (σ-ε) diagrams for the considered materials [16,17].

Some experimental data related to CVN energy is presented in Table 2. Charpy V-notch energy was also experimentally determined [16,17]. Experimentally-obtained results show that steel 1.4305 manifests a higher tensile strength and elongation than steel 1.7225, but CVN energy results show that the steel 1.4305 presents a lower CVN and lower fracture toughness than steel 1.7225, which can be attributed to the state of the as-received materials. Steel 1.7225 was a soft annealed as-received material (containing more percentage of carbon), compared to steel 1.4305, which was annealed and cold drawn (containing a lower percentage of carbon).

Table 2. Yield strength (σYS) and tensile strength (σTS) of the considered materials [16,17].

Material	σ_{YS} (Mpa)	σ_{TS} (Mpa)	CVN (J)	K_{Ic} (MPa· $m^{1/2}$)
1.4305	467	728	46	94.5
1.7225	415	617	166	212.1

In Figure 3, an optical micrograph of the as-received steel 1.4305 (soft annealed and cold drawn) is presented.

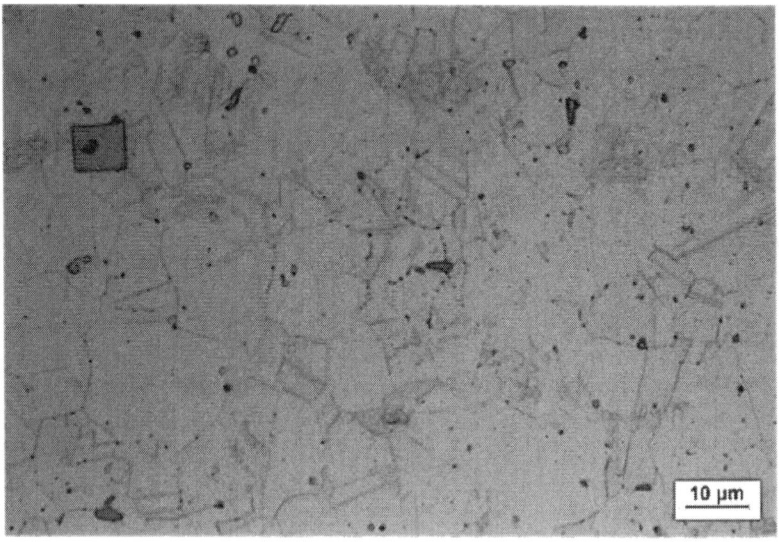

Figure 3. Optical micrograph of steel 1.4305; as-received material, soft annealed and cold drawn, cross-section of the specimen, aqua regia, 1000×.

In Figure 4, an optical micrograph of the as-received steel 1.7225 (soft annealed) is presented.

Considering the microstructure of steel 1.4305, it can be said that the basic microstructure of the as-received material is austenite, but there is also a mixture of austenite and ferrite. Considering the microstructure of steel 1.7225, it can be said that its main phase (main structure) consists of a thin pearlitic microstructure, where also a few ferrite and some particles of cementite can be observed.

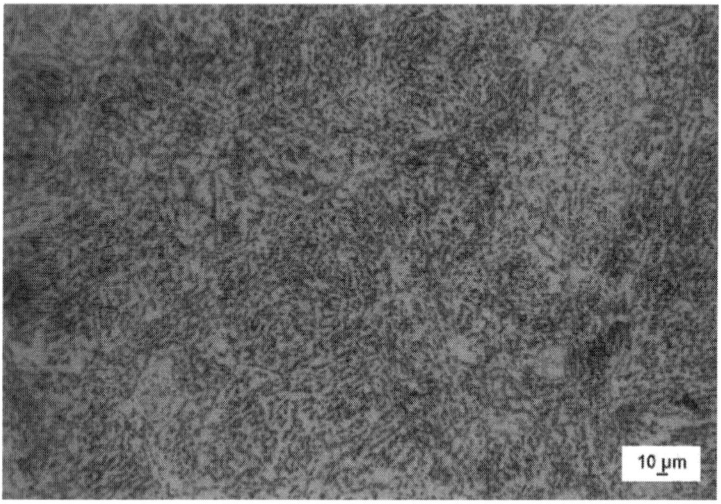

Figure 4. Optical micrograph of steel 1.7225; as-received material, soft annealed, cross-section of the specimen, 4% nital, 1000×.

2.2. Predicted Fracture Behavior of Considered Materials

J-integral is used to numerically predict the fracture behavior of the considered materials. J-integral was introduced by Rice and Cherepanov [19,20], separately, as a path-independent integral, which can be drawn around the tip of a crack and viewed both as an energy release rate parameter and a stress intensity parameter. In a two-dimensional form, it can be written as:

$$J = \int_{\Gamma} \left(w \mathrm{d}y - T_i \frac{\partial u_i}{\partial x} \mathrm{d}s \right) \quad (2)$$

where Ti = σijnj are components of the traction vector, ui are the displacement vector components, and ds is an incremental length along the arbitrary contour path Γ enclosing the crack tip.

In order to predict fracture behavior of steels 1.4305 and 1.7225, an experimental single specimen test method [15] following an elastic unloading compliance technique was numerically simulated. This test method uses measured crack mouth opening displacement to estimate the growing crack size. Resulting J-integral values can be taken as a fracture toughness parameter and plotted versus crack extension. The first step of the numerical procedure is to conduct a structural stress analysis.

According to the appropriate ASTM standard [15], 2D FE models of the two types of specimens, single edge notched bend (SENB), and disc compact tensile (DCT), are defined and initial a/W (W = 50 mm) ratios of 0.25, 0.5 and 0.75 are taken (Figure 5). The material behavior is considered to be a multilinear isotropic hardening type. FE models of specimens are meshed with 8-node isoparamateric quadrilateral elements. The mesh is refined around the crack tip in order to capture high deformation gradients in the regions where yielding occurs. To simulate compliance procedures of single specimen test method, quasi-static load was imposed on specimen. Only half of the specimen needs to be modeled due to their symmetry. The node releasing technique was used to simulate crack propagation.

FE stress analysis results taken from the integration points of finite elements surrounding the crack tip were used to evaluate J-integral values using the following equation [21]:

$$J = \sum_{p=1}^{np} W_p G_p \left(\xi_p, \eta_p\right) \tag{3}$$

where Wp is the Gauss weighting factor, np is the number of integration points, and Gp is the integrand evaluated at each Gauss point p:

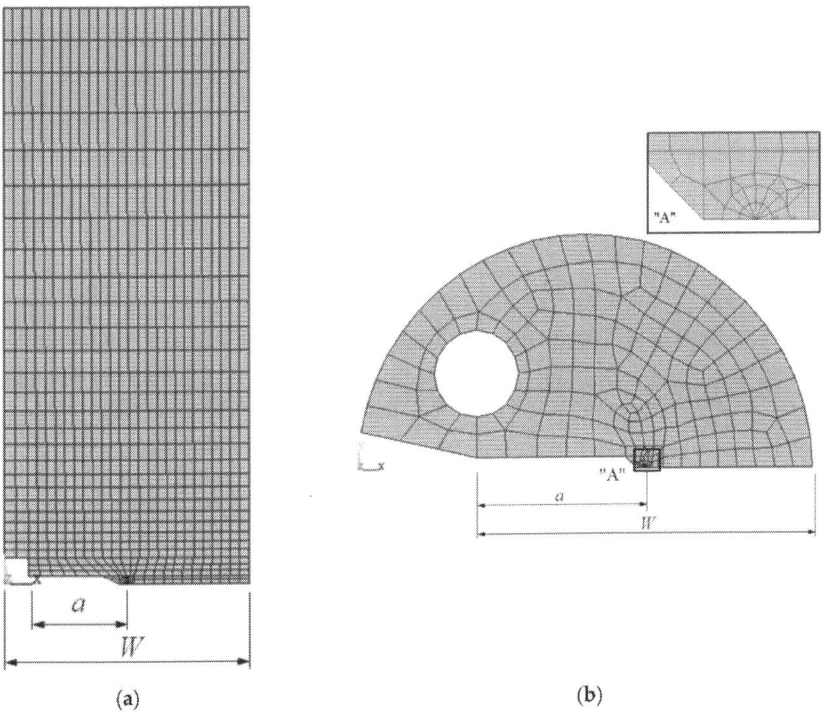

Figure 5. FE model of: (a) SENB specimen; (b) DCT specimen.

$$G_p = \left\{ \frac{1}{2} \left[\sigma_{xx} \frac{\partial u_x}{\partial x} + \sigma_{xy} \left(\frac{\partial u_x}{\partial y} + \frac{\partial u_y}{\partial x} \right) \frac{\partial u_x}{\partial x} + \sigma_{yy} \frac{\partial u_y}{\partial y} \right] \frac{\partial y}{\partial \eta} \right.$$
$$\left. \left[(\sigma_{xx} n_1 + \sigma_{xy} n_2) \frac{\partial u_x}{\partial x} + (\sigma_{xy} n_1 + \sigma_{yy} n_2) \frac{\partial u_y}{\partial x} \right] \sqrt{\left(\frac{\partial x}{\partial \eta} \right)^2 + \left(\frac{\partial y}{\partial \eta} \right)^2} \right\}_g \qquad (4)$$

J values are summed along the path that encloses the crack tip, giving total value of J. Three different paths around the crack tip have been defined in each example, and the average value was taken as a final. Although J-integral is independent of the chosen path, this was done in order to account for any possible J-value variations in the vicinity and away from the crack tip. This procedure had already been verified in cases when numerically-obtained parameters have been found to be corresponding with the experimental values [22].

Since no fracture experimental results were available for steels 1.4305 and 1.7225 to verify the accuracy of the procedure, J-integral values were first determined for the SENB specimen, with an initial crack length of a/W = 0.25, 0.5, 0.75, made of 1.6310 steel. Numerically-obtained results were compared with available experimental data for the same specimen configuration and material [23] (Figure 6). Good correspondence between numerically-predicted and experimental results encouraged further application of the J-integral calculation method.

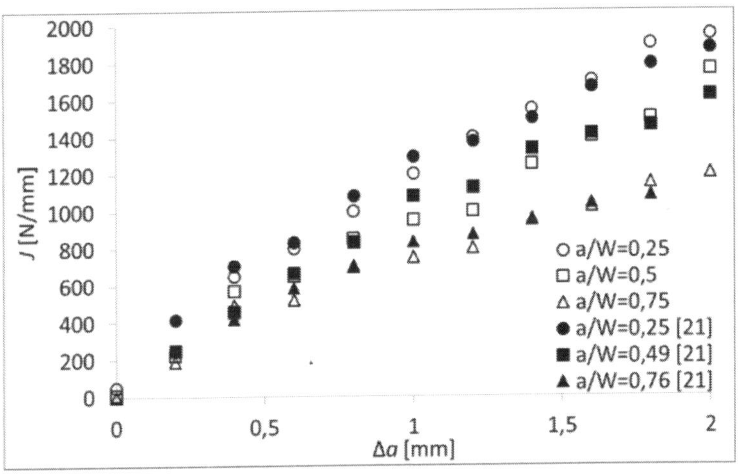

Figure 6. Comparison of numerically-predicted and experimentally-obtained J values for SENB specimens of 1.6310 steel.

3. RESULTS

The entire numerical procedure described in Section 2.2 was performed for FE models of SENB and DCT specimens. With respect to the geometry of specimens, initial a/W (crack length/width ratio with W = 50 mm) ratios of 0.25, 0.5 and 0.75 are taken with Δa = 0...2 mm (Δa being crack advance). For every case, J-integral values, as a measure of crack driving force, are calculated and the final results are shown in Figure 7 (for steel 1.4305) and Figure 8 (for steel 1.7225).

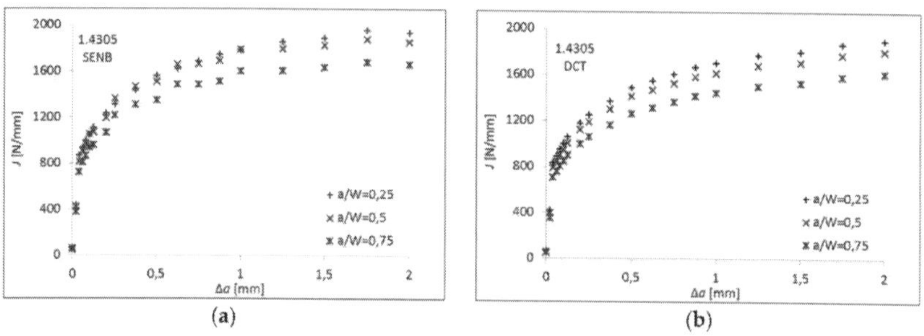

Figure 7. Numerically-predicted J values for steel 1.4305 using FE models of: (a) SENB specimen; (b) DCT specimen.

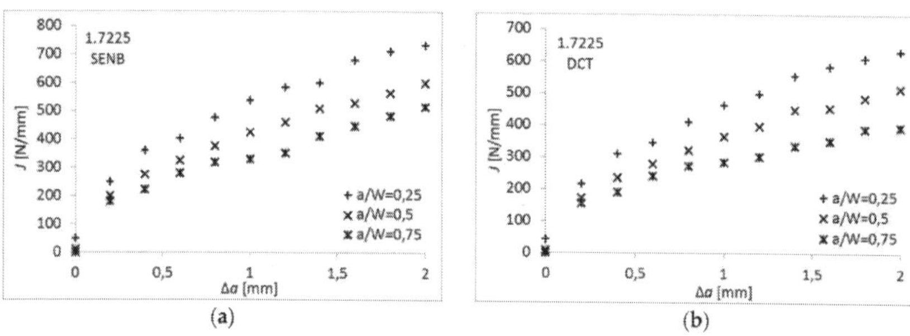

Figure 8. Numerically-predicted J values for steel 1.7225 using FE models of: (a) SENB specimen; (b) DCT specimen.

4. DISCUSSION

The fracture behaviors of the mentioned materials are given in Figure 7 and Figure 8, using J-integral as a measure of crack driving force. It can be noted that steel 1.7225 has higher resulting values of J-integral than steel 1.4305, making it more suitable for structures that need less susceptibility to fracture. The predicted difference in resistance to crack extension between steel 1.4305 and steel 1.7225 can be related to the different material properties and compositions (Table 1 and Table 2) of the two materials. Steel 1.4305 has a significantly higher value of nickel, which can contribute to the noted behavior. Nickel is usually added at over 8% (here 7.95%) content to chromium-nickel stainless steels in order to increase strength, impact strength, and toughness. It also improves resistance to oxidization and corrosion along with chromium. Chromium is added at over 12% content in stainless steels to significantly improve corrosion resistance. Added benefits are also hardenability, strength, response to heat treatment, and wear resistance.

In numerical modeling, crack geometry (a/W ratios) was kept the same for both materials in relative specimens, so the geometry probably could not contribute to a difference in J values for the two steels.

J-integral values differ for a/W = 0.75 when compared with a/W = 0.25 and 0.5, which are quite similar in terms of values for steel 1.4305. Additionally, higher a/W ratios correspond to lower J-integral values of materials and vice versa. J-integral values obtained using a DCT specimen FE model give somewhat conservative results when compared with ones obtained using the SENB specimen FE model. Experimentally-obtained KIc values and numerically predicted J values cannot be easily compared because J values go beyond the elastic behaviour of the material, where K is appropriate parameter. Further, J is taken as a measure of crack driving force here. Although the mentioned numerical procedure does not give results that can be directly related to ones obtained experimentally, given results can be useful for the assessment of fracture toughness.

The numerically-obtained data, along with experimentally-obtained yield strength, tensile strength, and CVN energy for steels 1.4305 and 1.7225 presented in this paper may be of importance for designers of engineering structures when concerning material selection. In addition, numerical assessment of J-integral could be useful as a predictor of the possible fracture behaviour of a material. Altogether, experimentally- and numerically-obtained data can be used in the design process to assess the possible load capacity of a structure.

ACKNOWLEDGMENTS

This work has been financially supported by the Croatian Science Foundation under the project 6876, and by the University of Rijeka under the projects 13.09.1.1.01 and 13.07.2.2.04. Funds for covering the costs to publish in open access publications have been provided.

AUTHOR CONTRIBUTIONS

Goran Vukelic and Josip Brnic conceived the idea of the research; Josip Brnic performed the experiments; Goran Vukelic analyzed the data; Goran Vukelic performed numerical analysis; Goran Vukelic and Josip Brnic wrote the paper.

REFERENCES

1. Fonte, M.; de Freitas, M. Marine main engine crankshaft failure analysis: A case study. Eng. Fail. Anal. 2009, 16, 1940–1947.
2. Zangeneh, Sh.; Ketabchi, M.; Kalaki, A. Fracture failure analysis of AISI 304L stainless steel shaft. Eng. Fail. Anal. 2014, 36, 155–165.
3. Moolwan, C.; Netpu, S. Failure analysis of a two high gearbox shaft. Procedia Soc. Behav. Sci. 2013, 88, 154–163.
4. Das, S.; Mukhopadhyay, G.; Bhattacharyya, S. Failure analysis of axle shaft of a fork lift. Case Stud. Eng. Fail. Anal. 2015, 3, 46–51.
5. Kossakowski, P.G. Simulation of ductile fracture of S235JR steel using computational cells with microstructurally-based length scales. J. Theor. Appl. Mech. 2012, 50, 589–607.
6. Dai, Q.; Zhou, C.; Peng, J.; He, X. Experiment, finite element analysis and EPRI solution for J-integral of commercially pure titanium. Rare Metal. Mat. Eng. 2014, 42, 257–263.
7. Huang, Y.; Zhou, W. J-CTOD relationship for clamped SE(T) specimens based on three-dimensional finite element analyses. Eng. Fract. Mech. 2014, 131, 643–655.
8. Koshima, T.; Okada, H. Three-dimensional J-integral evaluation for finite strain elastic-plastic solid using the quadratic tetrahedral finite element and automatic meshing methodology. Eng. Fract. Mech. 2015, 135, 34–63.
9. Saputra, A.A.; Birk, C.; Song, C. Computation of three-dimensional fracture parameters at interface cracks and notches by the scaled boundary finite element method. Eng. Fract. Mech. 2015, 148, 213–242. [Google Scholar] [CrossRef]
10. Zambrano, O.A.; Coronado, J.J.; Rodríguez, S.A. Failure analysis of a bridge crane shaft. Case Stud. Eng. Fail. Anal. 2014, 2, 25–32.

11. Fonte, M.; Duarte, P.; Anes, V.; Freitas, M.; Reis, L. On the assessment of fatigue life of marine diesel engine crankshafts. Eng. Fail. Anal. 2015, 56, 51–57.

12. Tawancy, H.M.; Al-Hadhrami, L.M. Failure of a rear axle shaft of an automobile due to improper heat treatment. J. Fail. Anal. Prev. 2013, 13, 353–358.

13. Bai, S.Z.; Hu, Y.P.; Zhang, H.L.; Zhou, S.W.; Jia, Y.J.; Li, G.X. Failure analysis of commercial vehicle crankshaft: A case study. Appl. Mech. Mater. 2012, 192, 78–82.

14. Fuller, R.W.; Ehrgott, J.Q., Jr.; Heard, W.F.; Robert, S.D.; Stinson, R.D.; Solanki, K.; Horstemeyer, M.F. Failure analysis of AISI 304 stainless steel shaft. Eng. Fail. Anal. 2008, 15, 835–846.

15. American Society for Testing and Materials (ASTM International). Metals Test Methods and Analytical Procedures; Annual BOOK of ASTM Standards; ASTM International: Baltimore, Maryland, MD, USA, 2005; Volume 03.01.

16. Brnic, J.; Turkalj, G.; Canadija, M.; Lanc, D.; Krscanski, S. Responses of austenitic stainless steel American iron and steel institute (AISI) 303 (1.4305) subjected to different environmental conditions. J. Test. Eval. 2012, 40, 319–328.

17. Brnic, J.; Turkalj, G.; Canadija, M.; Lanc, D.; Brcic, M. Study of the effects of high temperatures on the engineering properties of steel 42CrMo4. High Temp. Mater. Proc. 2015, 34, 27–34.

18. Roberts, R.; Newton, C. Interpretive Report on Small Scale Test Correlations with KIc Data; Welding Research Council Bulletin: New York, NY, USA, 1981; pp. 1–16.

19. Rice, J.R. A path independent integral and the approximate analysis of strain concentration by notches and cracks. J. Appl. Mech. 1968, 35, 379–386.

20. Cherepanov, G.P. The propagation of cracks in a continuous medium. J. Appl. Math. Mech. 1967, 31, 503–512.

21. Mohammadi, S. Extended finite element method; Blackwell Publishing: Singapore, 2008; pp. 56–58.

22. Vukelic, G.; Brnic, J. Prediction of fracture behavior of 20MnCr5 and S275JR steel based on numerical crack driving force assessment. J. Mater. Civ. Eng. 2015, 27, 14132–14132(6).

23. Narasaiah, N.; Tarafder, S.; Sivaprasad, S. Effect of crack depth on fracture toughness of 20MnMoNi55 pressure vessel steel. Mater. Sci. Eng. A 2010, 527, 2408–2411.

CHAPTER 7

Corrosion Resistance of API5L X52 Carbon Steel in Sulfide Polluted Environments

Randa Abdel-Karim*, Mohamed Ali Farag, Hafiz Abdel-Azim Ahmed, Saad El-Raghy

Department of Metallurgy, Faculty of Engineering, Cairo University, Giza, Egypt

ABSTRACT

In the present work, the corrosion behavior of base and welded API5L X-52 carbon steel coupons in synthetic ground water solution NS-4 was studied. Subsequently, the morphology and composition of the corrosion product deposits were determined using scanning electron microscopy (SEM). Based on weight loss as well as Tafel's measurements, increasing the sulfide ions content above 0.4 mM, had an inhibition effect on the corrosion behavior of carbon steel. According to EDAX analyses, increasing of sulfide concentration from 0.4 to 4 mM Na_2S, the Fe/S ratio decreased from 12.02 at 0.4 mM to 2.6 at 4 mM Na_2S.

KEYWORDS

Carbon Steel, Sulfides, Weight Loss, Anodic Dissolution, Polarization

1. INTRODUCTION

The corrosion of carbon steel in sulfide-containing solutions has attracted widespread attention for many years due to its importance in several industrial processes such as oil and gas production and transport, petroleum refining, and petrochemical/chemical processing [1] .

Some variables which affect the corrosion rate of steel in soil such as water, degree of aeration, pH, redox potential, resistivity, soluble ionic species, and microbiological activity. Concerning the environmental aspect, corrosion of steel structures in soil is a relevant problem due to soil and ground water contamination caused by failure of pipelines and other structures [2] .

The presence of some pollutants, mainly sulfide, in the oil can affect the performance of mild steel in the petrochemical industry, as this phenomenon is responsible for costly economic and human loss. Sulfide pollution of ground water can occur from the industrial waste discharge, biological and bacteriological process from seaweed, sulfide reduced bacteria, or industrial waste discharge [3] [4] .

Other applications where sulfide species have produced corrosion include water production, electric power, and marine applications. In addition, the corrosion of steel in CO_2, NaCl, Na_2CO_3, $NaHCO_3$, or Na_2S solutions constitutes the natural media often confronted in oil and gas prospecting and transportation areas [1] .

The effect of sulfide on the corrosion behavior of carbon steel in natural waters is studied by many authors [5] [6] . In the present work, the electrolyte used is a synthetic solution to simulate the soil or underground water conditions (NS-4 solution). This environment considers groundwater conditions similar to the pipelines that are exposed to underground environments while used for transmission lines. The corrosion behavior of steel in soil environment is studied using the weight loss, open circuit measurements (OCP) and Tafel's extrapolation analysis. Subsequently, the morphology and composition of the corrosion product deposits are determined using scanning electron microscopy (SEM) and energy dispersive spectroscopy (EDX).

2. EXPERIMENTAL

2.1. Materials

Experiments were conducted using API5L X-52 carbon steel pipes, provided by GASCO company-Egypt, with outer diameter 168.275 mm and the thickness 6.9 mm. The chemical composition and mechanical characteristics of this steel are given in Table 1 and Table 2. The hardness of the material was 155 HVN, and the microstructure was composed of pearlite and ferrite as shown in Figure 1. For corrosion measurements, rectangular base- as well as weld-coupons with dimensions 200 × 40 mm, were prepared and were subjected to grinding using Sic paper up to grid 1200 and polishing using alumina paste. In order to investigate the effect of plastic deformation, U-bend samples were also prepared with dimensions 200 × 40 × 6.9 mm. Three different diameters rolls of 30 mm, 40 mm, and 50 mm were used in the universal testing machine to give three different strains and hence three different stresses (Table 3).

Table 1. Chemical composition of API5LX-52 pipeline steel.

C	Si	P	S	Mn	Ni	Cr	Cu	Mo	V	Al
0.21	0.22	0.015	0.008	0.56	0.029	0.04	0.064	0.01	0.003	0.005

Table 2. Mechanical characteristics of API5LX-52 pipeline steel.

Specimen	W (mm)	Thickness (mm)	Yield Strength (Mpa)	UTS (Mpa)	Strain %
Weld	18.8	6.2	n.a	567	n.a
Base Metal	11.5	6.9	355	527	25

Figure 1. Microstructure of API5L X52 base metal (4" pipeline material).

2.2. Electrolytes

The synthetic underground water (NS-4) electrolyte used was near-neutral pH environment (pH 8.4), whose chemical composition is given in Table 4 after Parkins et al. [7] . The sulfide stock solution was prepared by dissolving ($Na_2S.9\ H_2O$) crystals in NS-4 solution. The effect of sulfide ions concentration on the corrosion behavior of carbon steel was studied in NS-4 solution containing 0.4, 2, 4, and 10 mM Na_2S with conductivity values 1456, 2230, 2680 and 5030 μS, respectively.

2.3. Immersion Test

The corrosion performance of carbon steel coupons in synthetic underground water (NS-4), contaminated with different sulfide contents, was studied by measuring the weight loss as well as the open circuit potential (OCP) at time intervals of 7 days for 3 month. Samples were rinsed with distilled water and carefully dried before taking weight change measurements. The corresponding corrosion rates (CR) in mils per year (mpy), based on weight loss measurements, were calculated using the following equation [8] :

$$CR = 534\frac{W}{DAt} \tag{1}$$

where W is the weight loss in gm (g), D is the density of steel g/cm^3, A is the total surface area (cm^2) and t is the total exposure time in hours.

2.4. Potentiodynamic Test

The potentiodynamic test was conducted using a standard three electrodes cell. Each specimen was used as a working electrode with an exposure area of 1 cm^2. A saturated calomel electrode was used as a reference electrode and the counter electrode was a platinum electrode. The cell was connected to PARSATAT4000 potentiostat. The scanning range was set from −200 mV to +200 mV with respect to open circuit potential using a scanning rate 0.5 mV/sec. Versa Studio Software was used for drawing polarization curves, as well as Tafel's plot.

2.5. Surface Characterization

SEM (JEOL JXA-840A) was used to demonstrate the microstructure and surface morphology of the corrosion products. Local compositions were obtained using the energy dispersive spectroscopy (EDS) unit attached to the SEM.

Table 3. Roll Diameter vs. strain & stress obtained.

Roll Diameter	Strain	stress
30 mm	23%	158%SMYS
40 mm	17.25%	156%SMYS
50 mm	13.8%	152%SMYS

Table 4. Chemical composition of NS-4 synthetic ground water solution [8] .

Chemicals	g/l
$NaHCO_3$	0.483
KCl	0.122
$CaCl_2$	0.093
$MgSO_4$	0.131

3. RESULTS

3.1. Open Circuit Potential (OCP)

Figures 2-4 summarize the potential measurements (OCP) for both base and weld metal specimens after immersion for 12 weeks in both working conditions (stressed and unstressed samples) as a function of sulfide ions content. It is concluded that the OCP values for stressed samples were more negative than the unstressed ones. The OCP values ranged from −680 mV up to −730 mV. There was a slight change in potential values by increasing the sulfide content, where the OCP values were in the range of ~ −680 to −700 mV. The most negative OCP was in case of NS-4 solution containing 0.4 mM Na_2S, as shown in Figure 4.

3.2. Weight Loss Measurements

Figure 5 illustrates a summary of corrosion rate (mpy) based on weight loss measurements for base as well as weld metal coupons after immersion for 3 month, as a function of sulfide content. For base metal coupons, the highest corrosion rate was obtained in NS-4 + 0.4 mM Na_2S (0.09 mpy). Increasing the sulfide ions content above 0.4 mM, had an inhibition effect on the corrosion behavior of carbon steel. The lowest corrosion rate (0.06 mpy) was obtained in NS-4 + 4 mM Na_2S. Similarly, for weld metal coupons, the highest corrosion rate (0.18 mpy) was obtained in NS-4 + 0.4 mM Na_2S. Weld metal coupons showed higher corrosion rate than its correspondents of base metal immersed in the same solutions.

3.3. Potentiodynamic Test

The electrochemical parameters calculated using the Tafel plots (Figure 6) confirmed that the addition of sulfide increases the corrosion current, i_{corr} and consequently the corrosion rates (mpy). The anodic branches present a uniform dissolution for examined specimens in all solutions.

Corrosion data obtained from Tafel's plot are summarized in Table 5. From this table, the extent of increase in i_{corr} and corrosion rates (mpy) is found to be a function of the concentration of Na_2S. The addition of Na_2S has an inhibition effect of the corrosion behavior of carbon steel samples. This trend is in agreement with test result based on weight loss measurements. Also from Table 5, and in agreement with the OPC measurement, the most negative corrosion potential (−768 mV) was obtained in NS-4 solution containing 4 mM Na_2S.

Corrosion rate (mpy) calculated from the Tafel's plots are presented in Figure 7. According to Figure 7, the weld metal specimen had higher corrosion rate than base metal one for both working conditions (stressed and unstressed). In case of stressed welded samples, the corrosion rates were 18, 32, and 15 mpy for NS-4 solutions with 0, 0.4, and 4 mM Na_2S, respectively.

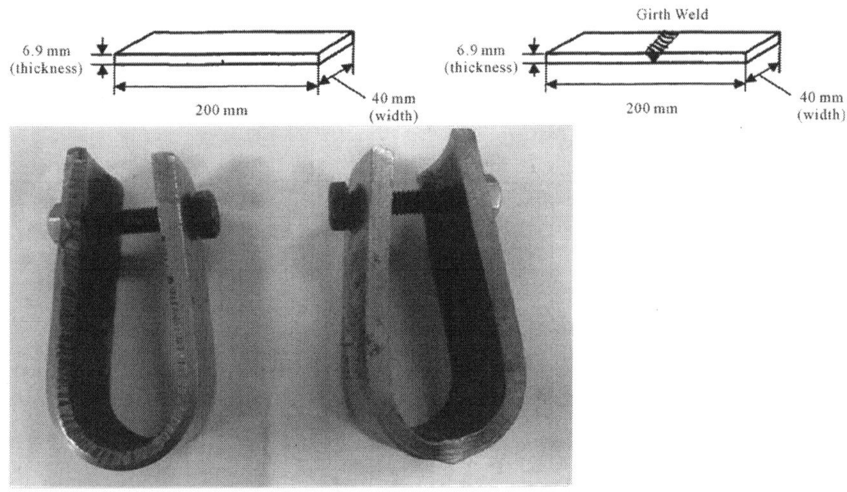

Figure 2. Test Specimens. (a) Base-metalpecimen; (b) Weld Specimen; (c) U-bend stressed specimens.

3.4. SEM Surface Morphology and EDAX Analysis

The SEM images of base as well as weld metal coupons after immersion for three month in different sulfide polluted environments are illustrated in Figure 8. The SEM micrograph of base metal coupon which was immersed in NS4 solution, is characterized by dense layer (Figure 8(a)). From Table 6, EDAX analyses showed a high content of iron (88.14%) and the presence of oxygen (6.05%), carbon (4.58%), and manganese (%1.23). On the other hand, the SEM micrograph of weld metal coupon (Figure 8(b)) is characterized by flowery shaped layer formed on the metal surface immersed in the same above environment.

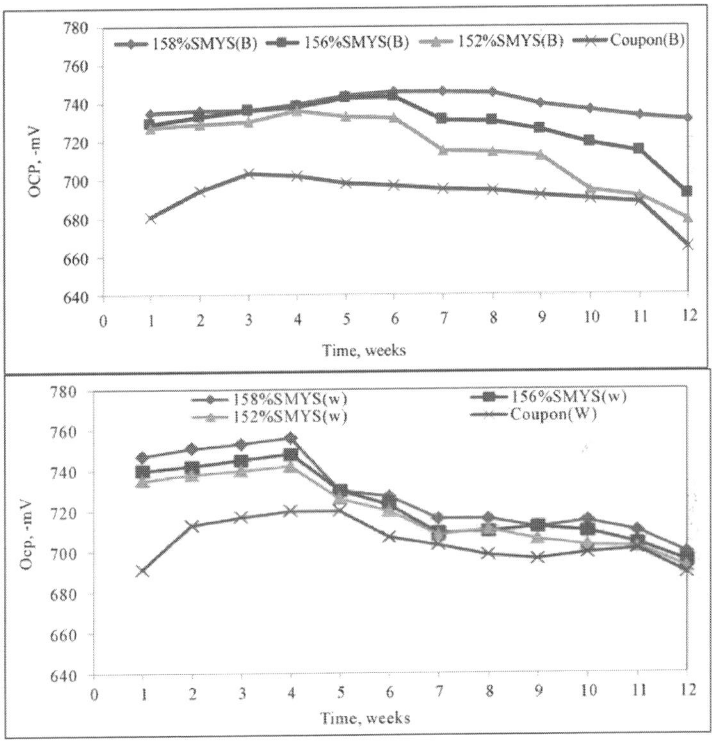

Figure 3. Effect of plastic deformation applied on OCP measurements in NS-4 solution. (a) Base-metal; (b) Welded-metal.

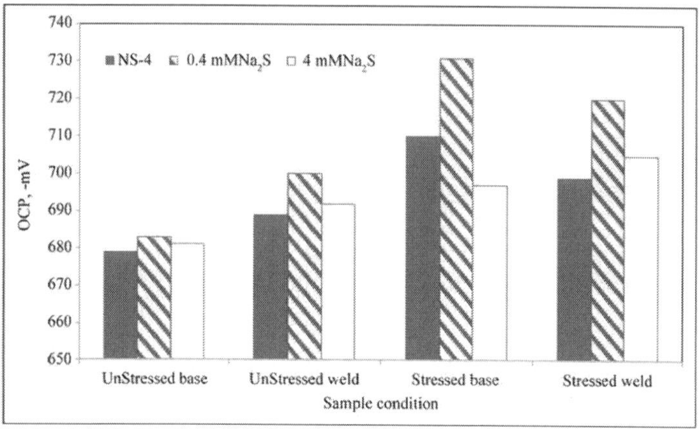

Figure 4. Open circuit potential of specimens for different sulfide-contaminated environments and different samples preparation conditions after immersion for 12 weeks.

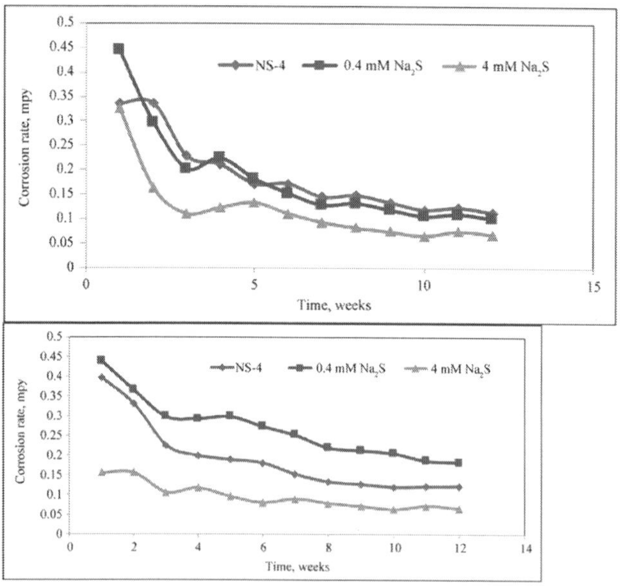

Figure 5. Corrosion rates (mpy) of welded-coupons after immersion in different sulfide- contaminated environments for 12 weeks-based on weight loss measurements. (a) Base metal; (b) Welded metal.

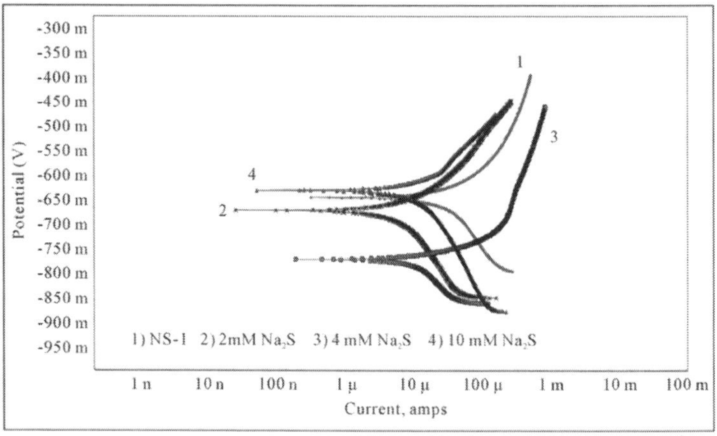

Figure 6. Polarization curves of weld coupons measured in sulfide contaminated environments.

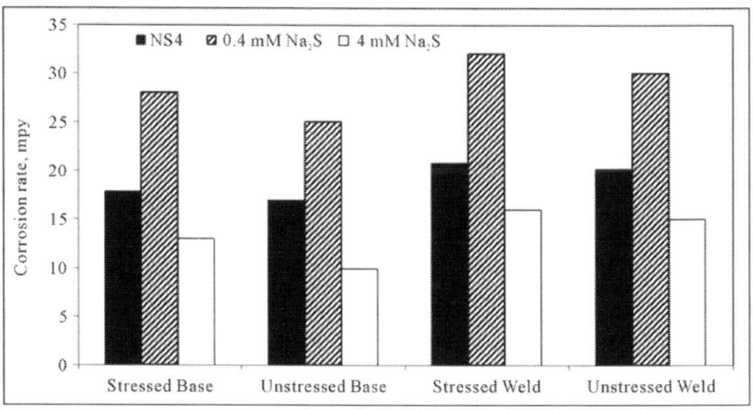

Figure 7. Corrosion rates of carbon steel specimens as a function of sulfide content as well as stresses applied.

Table 5. Corrosion data of API5LX-52 welded coupons immersed in different sulfide polluted electrolytes.

Condition	I corr ($A/\mu cm^2$)	Corrosion rate (mils/year)	Ecorr (mV)	βa mv/decade	βc mv/decade
NS4	21.2	20.2	−642	130	160
NS4 + 0.4 MNa_2S	31	30	−720	162	170
NS4 + 2 mNa_2S	15.34	14.74	−667	151.5	200
NS4 + 4 mNa_2S	15.5	15	−768	37.4	162.8
NS4 + 10 mNa_2S	18.8	18.1	−627	180	175

Figure 8(c) shows the SEM micrograph of weld metal coupon which was immersed in NS-4 + 0.4 mM Na_2S after 3 months. The SEM image shows a thin porous layer, and consequently non protective layer. EDAX analysis of this layer showed a lower content of iron (50.1%) and the presence of higher content of oxygen (32.7%), carbon (9.08%), magnesium (3.81%) and sulfur (4.17%), by increasing the sulfide content.

Figure 8(d) shows the SEM micrograph of the surface of the weld metal coupon which was immersed in NS-4 + 2 mM Na_2S for 3 months. The SEM image shows a dense spongy layer formed on the metal surface.

Figure 8(e) shows the SEM micrograph weld metal coupon which was immersed in NS-4 + 4 mM Na_2S for 3 months. The SEM image shows a dense layer with some bright particles formed on the metal surface. EDAX analysis (Table 6) indicated a lower iron content (41.47%) and the presence of of oxygen (27.9%), carbon (9.05%), magnesium (4.54%) and higher sulfur content (15.71%), in case of 4 mM Na_2S.

Figure 8(f) shows the SEM micrograph of the surface of the weld metal coupon which was immersed in NS-4 + 10 mM Na_2S for 3 months. The SEM image shows a dense protective layer with many bright particles formed on the metal surface.

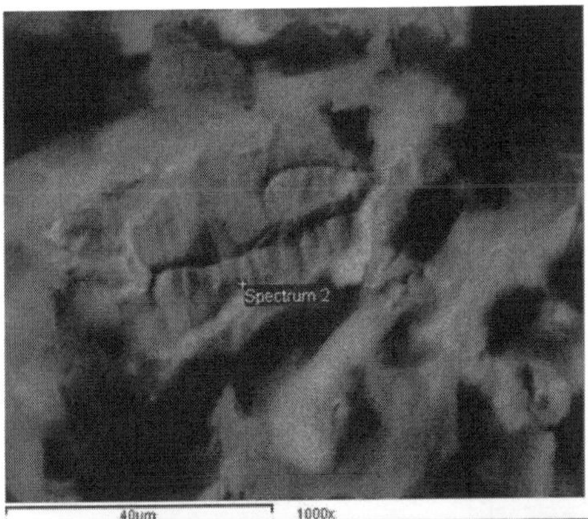
Spectrum 2
40µm
1000x

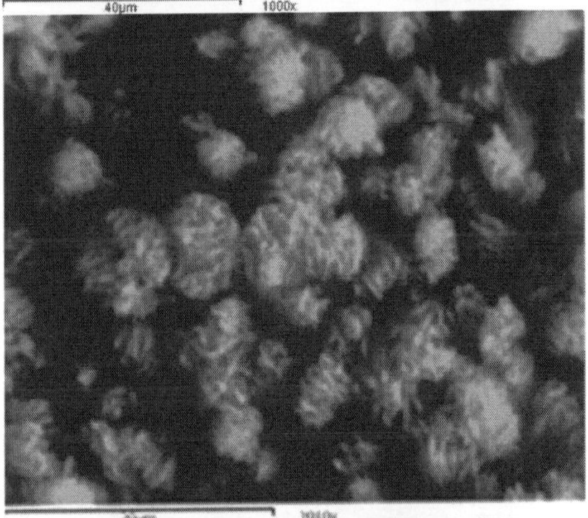

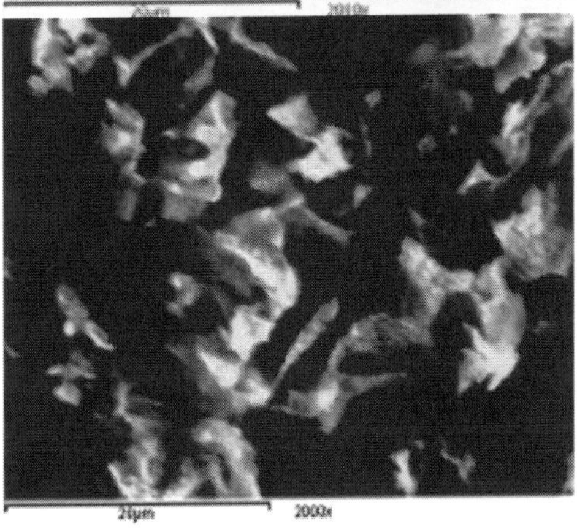
20µm
2000x

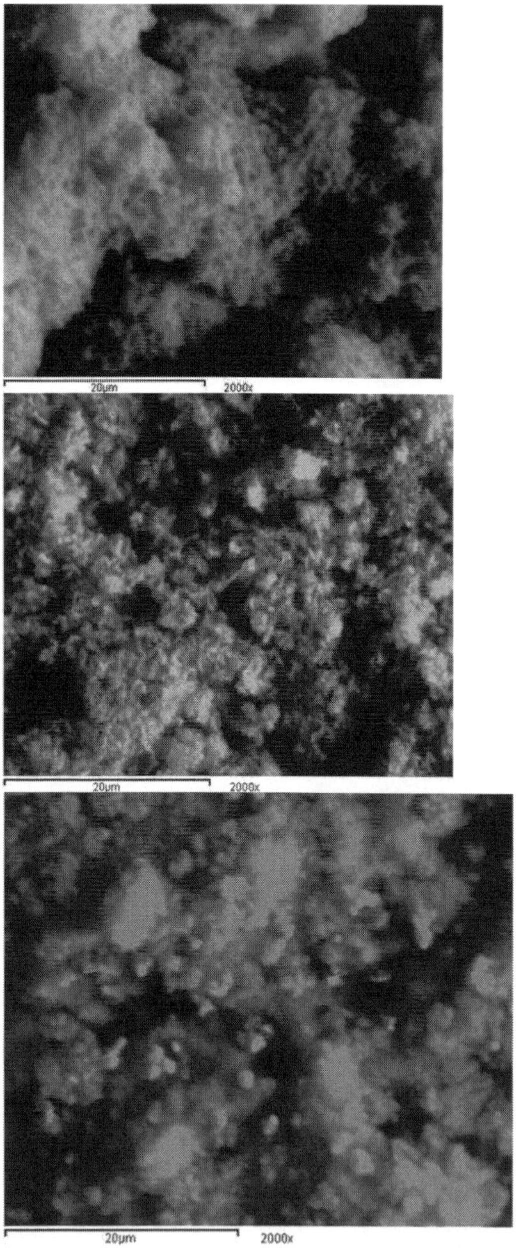

Figure 8. SEM of weld-metal coupons after immersion in sulfide contaminated environments for 3 month.

Table 6. EDAX analysis of API5LX-52 coupons taken at the weld zones after 3 month of immersion.

Condition	% C	% O	% Mn	% Mg	% S	% Si	% Fe
NS-4	4.58	6.05	1.23	0.0	0.0	0.0	88.14
NS-4 + 0.4mM Na_2S	9.08	32.79	0.0	3.81	4.17	0.0	50.15
NS-4 + 4 mM Na_2S	9.05	27.90	0.0	4.54	15.71	0.0	41.47

4. DISCUSSION

According to this work, and from potentiodynamic polarization test, it was noticed that stressed specimens showed higher corrosion rate than unstressed coupons in both cases of base and weld metal specimens. This is in agreement with other study showing that the corrosion current increased significantly and monotonically for samples with increasing plastic deformation [9] . The value of applied strain was based on the following formula;

$$\varepsilon = t/2R \text{ when } t < R \tag{2}$$

where ε is the applied strain, t is the specimen thickness and R is the radius of the bend [10] .

When steels are plastically deformed, some fraction of the deformation energy (~5%), is retained internally. The major portion of this stored energy is as strain energy associated with dislocations. The increase in the amount of strain energy would allow easier dissolution on specific sites since this energy is released on dissolution. This causes an increase in the corrosion rate with an increase in the amount of cold work [11] .

It was also recognized from potentiodynamic polarization test, and in agreement with weight loss test, that weld metal specimen suffered higher corrosion rate than base metal ones. This can be arisen from the fact that weld-metal composition (which is normally optimized for mechanical properties) tend to be slightly anodic to the parent steel in certain environments. Therefore, the weld metal corrodes at a higher rate than the base metal [12] [13] .

The deterioration of metal due to contact with hydrogen sulfide (H_2S) and moisture is called sour corrosion which is the most damaging to drill pipe. Although H_2S is not corrosive by itself, it becomes a severely corrosive agent in the presence of water, leading to pipeline embrittlement. Hydrogen sulfide when dissolved in water is a weak acid, and therefore, it is a source of hydrogen ions and becomes corrosive. The corrosion products are iron sulfides (FeS_x) and hydrogen. Iron sulfide forms a scale that at low temperature can act as a barrier to slow corrosion. The forms of sour corrosion are uniform, pitting, and stepwise cracking. The general equation of sour corrosion can be expressed as follows [14] :

$$H_2S + Fe + H_2O \rightarrow FeS_x + 2H + H_2O \quad (3)$$

There is a possibility for formation of H_2S in the presence of some sulfide salts, according to the following equations [15] [16] ;

$$Na_2S \rightleftarrows Na^+ + S^{-2} \quad (4)$$

$$S^{-2} + H_2O \rightleftarrows HS^- + OH^- \quad (5)$$

$$HS^- + H_2O \rightleftarrows H_2S + H_2O \quad (6)$$

Another possible mechanism for iron dissolution in aqueous electrolytes containing H_2S is based on the formation of a tetragonal ferrous Sulfide (mackinawite) FeS as follows;

$$Fe + H_2S + H_2O \rightleftarrows FeHS - ads + H_3O^- \quad (7)$$

$$FeHS\text{-}ads \rightleftarrows FeHS^+ + 2e^- \quad (8)$$

$$FeHS^+ + H_3O^+ \rightleftarrows Fe + H_2S + H_2O \quad (9)$$

$$Fe^{+2} + HS^- \rightleftarrows FeS + H^+ \quad (10)$$

The FeS layer of steel is not stable, it is removed from the steel surface in an acidic environment, forming again H_2S, enhancing corrosion:

$$FeS + 2H \rightleftarrows Fe^{2+} + H_2S \quad (11)$$

There is a possibility that the surface of the steel will be covered by Fe_3S_4 protective layer that blocks the carbon steel from further corrosion. This layer could be doped with other corrosion products:

$$3FeS + H_2S \text{ ads} \rightarrow Fe_3S_4 + 2H^+ + 2e^- \tag{12}$$

Subsequently

$$3Fe^{2+} + 3S^{-2} + H_2S \rightarrow Fe_3S_4 + 2H^+ + 2e^- \tag{13}$$

This is in agreement with the study mentioned in the report from ICPTT [16] , and explains the reduction in corrosion rate measured by weight loss as well as Tafel's extrapolation technique, by increasing the sulfide ions content.

Increasing the sulfide ions content above 0.4 mM, had an inhibition effect on the corrosion behavior of carbon steel. For base metal coupons, the highest corrosion rate was obtained in NS-4 + 0.4 mM Na_2S (31 mpy). The lowest corrosion rate was obtained by conducting potentiodynamic test in NS4 + 4 mM Na_2S (~15 mpy).

The weight loss test of coupons immersed in NS4 + 4 mM Na_2S agreed with these results obtained from potentiodynamic test, which showed a low corrosion rate at high concentration of Na_2S. The reason for this low corrosion rate may be because of the use of sodium sulfide as the source of H_2S. At higher pH sodium sulfide may not produce H_2S gas and the dominant species could be S^{2-}, and the absence of H^+ ions decreases the corrosion rate. Also the reaction product, formed at higher pH is metal sulfides, which when precipitates on the surface of the metal gives protection to the metal as these metal precipitates have extremely low solubility. This is in agreement with a study performed by M. Koteeswaran [17] on CO_2 and H_2S corrosion in oil pipe lines.

In addition, some results demonstrated that the presence of hydrogen sulfide can inhibit the corrosion of iron under certain special conditions due to the formation of a ferrous sulfide—pyrite protective film [18] -[20] .

Recently, Sherar et al. [20] investigated the effect of inorganic sulfide on carbon steel corrosion in a solution containing chloride, bicarbonate and sulfate (pH 8.9) by following the evaluation of corrosion potential (Ecorr) and periodically measuring the polarization resistance (R_P) over an exposure period of a few months. When freshly-polished carbon steel was exposed directly to sulfide a low corrosion rate was observed. However, when sulfide was added to pre-corroded steel, the corrosion rate tripled [21] . These observations were consistent with the results reported by Newman et al. [21] , who hypothesized that a pre-corroded surface prevents FeS passi- vation.

The SEM image of the weld coupon immersed in NS-4 + 0.4 mM Na_2S showed a thin porous layer which is not protective formed on the steel's surface, and this is in support of high corrosion rate in this solution. The SEM image of the weld coupon at high concentrations of Na_2S showed a compact passive corrosion product film formed on the steel surface. The EDAX analysis showed that in case of NS-4 electrolyte, corrosion product formed on the metal surface was composed of; 4.58% C, 6.05% O, 1.23% Mn, 88.14% Fe. The protective product formed on the metal surface in case of 4 mM Na_2S is composed of; 9.08% C, 32.79% O, 3.81% Mg, 4.17% S, 50.15% Fe. It was also noticed that with the increase of sulfide concentration from 0.4 to 4 mM, the Fe/S ratio decreased from 12.02 at 0.4 mM, to

2.63 at 4 mM Na_2S. This shows that increasing the sulfide concentration leads to the formation of sulfur-rich iron sulfide [5] .

5. CONCLUSIONS

1) Increasing the sulfide ions content above 0.4 mM Na_2S, had an inhibition effect on the corrosion behavior of carbon steel, due to formation of protective film.

2) There was a slight change in potential values by increasing the sulfide content.

3) The protective product formed on the metal surface in case of NS-4 + 4 mM Na_2S was composed of; 9.08% C, 32.79% O, 3.81% Mg, 4.17% S, 50.15% Fe.

4) Increasing of sulfide concentration from 0.4 to 4 mM, the Fe/S ratio decreased from 12.02 at 0.4 mM to 2.6 at 4 mM. This showed that increasing the sulfide concentration led to the formation of sulfur-rich iron sulfide.

5) The open circuit potential values for stressed steels were more negative than the unstressed samples.

6) Weldments of carbon steel showed more tendency for corrosion than base steel.

ACKNOWLEDGEMENTS

This paper is extracted from a thesis submitted to the faculty of Engineering-Cairo University as partial fulfillment of the M.Sc. degree by one of the authors in 2015 (Eng. Mohamed Ali Farag).

REFERENCES

1. Mendili, Y.E., Abdelouas, A. and Bardeau, J. (2014) The Corrosion Behavior of Carbon Steel in Sulfide Aqueous Media at 30 ?C, JMEPEG, 23, 1350-1357.
2. Hamdy, A., Shoeib, M.A, Saehand, A.G. and Barakat, Y. (2008) The Electrochemical Behavior of Mild Steel in Sulfide Polluted NaCl at Different Velocities. International Journal of Electrochemical Science, 3, 1142-1148.
3. de Freitas, V., Lins, C., Leonard, M., Ferreira, M. and Alves Saliba, P. (2012) Corrosion Resistance of API X52 Carbon Steel in Soil Environment. Journal of Materials Research and Technology, 1, 161-166.
4. Ashour, E.A., Khorshed, L.A., Youssef, G.I., Zakria, H.M. and Khalifa, T.A. (2014) Electrochemical and Stress Corrosion Cracking Behavior of Alpha-Al Bronze in Sulfide-Polluted Salt Water: Effect of Environmentally-Friendly Additives. Materials Sciences and Applications, 5, 10-19.
5. Xuejun, X., Wanqin, Y., Shunan, C., Ling, P., Xunjie, G. and Keru, P. (2005) The Effect of Sulfide on the Corrosion Behavior of Carbon Steel in Natural Waters. Anti-Corrosion Methods and Materials, 52, 145-147.
6. Fouda, A.S., El-Khateeb, A.T., Ibrahim, M. and Fakih, M. (2015) Adhatoda Aqueous Plant Extract as Environmentally Being Corrosion Inhibitor for Carbon Steel in Sanitation Water in Polluted NaCl Solutions and Its Biological Effect on Bacteria. Nature and Science, 13, 71-82.
7. Parkins, R.N., Blanchard, W.K.J. and Delanty, B.S. (1994) Transgranular SCC of High Pressure Pipelines near Neutral pH. Corrosion, 50, 394-408.

8. Ovri, J.E.O., Okeahailam, S.I. and Onyemaobi, O.O. (2013) Microbial Corrosion of Mild and Medium Carbon Steel. Journal of Engineering Science and Technology, 8, 639-653.

9. Li, X.C., Eadie, R.L. and Luo, J.L. (2008) Influence of Plasticity on Corrosion and Stress Corrosion Cracking Behavior in near Neutral pH Environment. 43, 297-303.

10. (2009) ASTMG30, Standard Practice for Making and Using U-Bend Stress-Corrosion Test Specimens.

11. Hertzberg, R.W. (1996) Deformation and Fracture Mechanics of Engineering Materials. 4th Edition, John Wiley & Sons, Inc., Hoboken.

12. Lee, C.-M., Bond, S. and Woollin, P. (2005) Preferential Weld Corrosion, Effects of Weldment Microstructure and Composition. Proceedings of Corrosion 2005, NACE International, Paper 05277.

13. Popoola, L.T., Grema, A.S., Latinwo, G.K., Gutti, B. and Balogun, A.S. (2013) Corrosion Problems During Oil and Gas Production and Its Mitigation. International Journal of Industrial Chemistry, 4, 1-15.

14. Yuan, G., Wang, S., Zhang, Y. and Huang, Y. (2009) Role of Sulfide Ion on Corrosion Performance of Pipeline Steel in Simulated Inner-Seeper Environment. International Conference on Pipelines and Trenchless Technology ICPTT 2009, Shanghai, 18-21 October 2009, 1643-1651.

15. Tang, J., Shao, Y., Zhang, T., Meng, G. and Wang, F. (2011) Corrosion Behavior of Carbon Steel in Different Concentrations of HCI Solutions Containing H2S at 90 ?C. Corrosion Science, 53, 1715-1723.

16. Koteeswaran, M. (2010) CO2 and H2S Corrosion in Oil Pipelines. M.Sc. Thesis, Faculty of Mathematics and Natural Science, University of Stavnger, Norway.

17. Arenas-Marteniz, L.F. (2012) Hydrogen Sulfide Corrosion of Weld Regions in API X52 Steel. Ingeniria, Investigacion y Tecnologia, 13, 473-478.

18. Ma, H., Cheng, X., Chen, S., Wang, C., Zhang, J. and Yang, H. (1998) An AC Impedance Study of the Anodic Dissolution of Iron in Sulfuric Acid Solutions Containing Hydrogen Sulfide. Journal of Electroanalytical Chemistry, 451, 11-17.

19. Ma, H., Cheng, X., Li, G., Chen, S., Quan, Z., Zhao, S. and Niu, L. (2000) The Influence of Hydrogen Sulfide on Corrosion of Iron Under Different Conditions. Corrosion Science, 42, 1669-1683.

20. Sherar, B.W.A., Keech, P.G., Noel, J.J. and Shoesmith, D.W. (2013) The Effect of Sulphide on Carbon Steel Corrosion Behavior in Anaerobic Near-Neutral Saline Solutions. Corrosion, 1, 67-76.

21. Newman, R.C., Rumash, K. and Webster, B.J. (1992) The Effect of Pre-Corrosion on the Corrosion Rate of Steel in Neutral Solutions Containing Sulfide: Relevance to Microbially Influenced Corrosion. Corrosion Science, 33, 1877-1884.

CHAPTER 8

Corrosion Behavior in 3.5% NaCl Solutions of γ-TiAl Processed by Electron Beam Melting Process

Asiful Hossain Seikh [1], Ashfaq Mohammad [2,*], El-Sayed M. Sherif [1,3,*] and Abdulrahaman Al-Ahmari [1,2]

[1] Centre of Excellence for Research in Engineering Materials, Advanced Manufacturing Institute, King Saud University, Riyadh 11421, Saudi Arabia

[2] Princess Fatima Alnijiris's Research Chair for Advanced Manufacturing Technology (FARCAMT), Advanced Manufacturing Institute, King Saud University, Riyadh 11421, Saudi Arabia

[3] Electrochemistry and Corrosion Laboratory, Physical Chemistry Department, National Research Centre, Cairo 12622, Egypt

ABSTRACT

In this work, the corrosion behavior of γ-TiAl alloy produced by electron beam melting (EBM) process in 3.5% NaCl solution was reported. The study has been performed using potentiodynamic polarization resistance and electrochemical impedance spectroscopy techniques and complemented by scanning electron microscopy investigations. All measurements were carried out after different periods of alloy exposure in the chloride solutions and at different temperatures. The results showed that the EBM produced γ-TiAl alloy has excellent corrosion resistance confirmed by the high values of polarization resistance and the low values of corrosion current and corrosion rate. With increase in immersion time, the corrosion potential moved to a higher positive value with a decrease in corrosion current and corrosion rate, which suggests an improvement in corrosion resistance. On the other hand, the increase of temperature was found to significantly increase the corrosion of the processed γ-TiAl alloy.

KEYWORDS

titanium aluminides; corrosion; electron beam melting; polarization; EIS

1. INTRODUCTION

Presently, there has been an increasing demand in the development of intermetallic alloys due to the stringent requirements of using high strength and low weight alloys in the automobile and modern aircraft industries. Gamma titanium aluminides (γ-TiAl) are intermetallic alloys, proposed to be used in advanced aircraft turbines [1,2,3]. γ-TiAl has better high temperature properties and higher specific strength than the conventional superalloys used in turbine fabrication. In addition, the use of γ-TiAl is preferred in aero-turbine industry because of its other several features. They possess light weight and offer superior creep, oxidation, and burn resistance than other titanium alloys. These alloys also have enhanced strength at elevated temperatures. The incorporation of several alloying elements like Al, Nb, Mo, Cr etc. enhance the mechanical properties as well as corrosion resistance properties for titanium based alloys due to the formation of protective oxide layers [4,5]. For that it goes without saying that γ-TiAl should also have a good resistance to the harsh corrosive environment inside the turbine.

γ-TiAl alloys having intermetallic properties demonstrate limited plasticity. Manufacturing highly complex geometries from such a material is a quite challenging exercise. Reports are available on processing these alloys through hot forging [6], machining [7], and powder metallurgy [8] routes. It can be deduced from these reports that, these technologies require multiple pre- and post-processing steps. Added to this is the very high cost. A slightly better route is casting. Nevertheless, casting has its own problems such as poor fluidity, easy cracking at stress concentrated portions [9].

Considering the limitations of conventional manufacturing routes, a need exists to look for other options. Recently, electron beam melting (EBM) was shown to produce parts having high strength from γ-TiAl [1]. Yet, it is not well known, how EBM processing would affect properties such as corrosion resistance of the actual γ-TiAl material. Due to the major application in military aircraft, exposure of these materials to marine environment leads major corrosion problem due to chloride ions presents. To date, there is little data available on the corrosion resistance properties of TiAl alloys. Delgado et al. [10] reported the corrosion behavior of γ-TiAl in Ringer's solution. It has been found that γ-TiAl has very good corrosion resistance similar to Ti-6Al-4V. Gurappa [11] has investigated the degradation of Ti-24Al-15Nb alloys under different environmental conditions and reported that a protective oxide layer is formed on its surface in marine and industrial environments.

The present study aims at investigating the corrosion resistance of γ-TiAl alloy that was processed using EBM technique after different exposure periods of time, namely, 0, 1, 4, and 24 h in stagnant aerated 3.5% NaCl solutions. It is known that 3.5% NaCl solution provides a simulated marine environment, which is the most corrosive medium for these alloys. The work was carried out at room temperature using potentiodynamic polarization and electrochemical impedance spectroscopy (EIS) measurements, in addition to the use of laser diffraction and scanning electron microscopy in the characterization of the γ-TiAl alloy.

2. EXPERIMENTAL PROCEDURE

2.1. Production of Ti Aluminide Alloy

γ-TiAl powder with nominal composition of Ti-48Al-2Cr-2Nb (at. %) supplied by Arcam AB (Mölndal, Sweden) was used as feedstock material. The powder particles were found to be in the range 45–180 μm with a mean of around 110 μm. This was measured by the laser diffraction technique as depicted in Figure 1. Moreover, the morphology of the powder particle was more or less spherical with smaller satellite particles adhering to the bigger ones, as can be seen from the SEM image of Figure 2.

EBM experiments were done on A2 model supplied by ARCAM AB. More details about EBM process setup can be found elsewhere [1]. The specimen dimensions were 25 × 25 × 13 mm height. A 10 mm thick square (100 × 100 mm) stainless steel plate was chosen as the substrate upon which the parts would be built. All the parts after being saved in STL format in Magics (Version 17; Materialise NV, Leuven, Belgium), were then transferred to Build Assembler software (Version 3; Arcam AB, Mölndal, Sweden). This particular software converts the STL geometry into thin slices and writes the output as machine specific ABF file. The ABF file instructs the machine control on how to proceed with the actual part production.

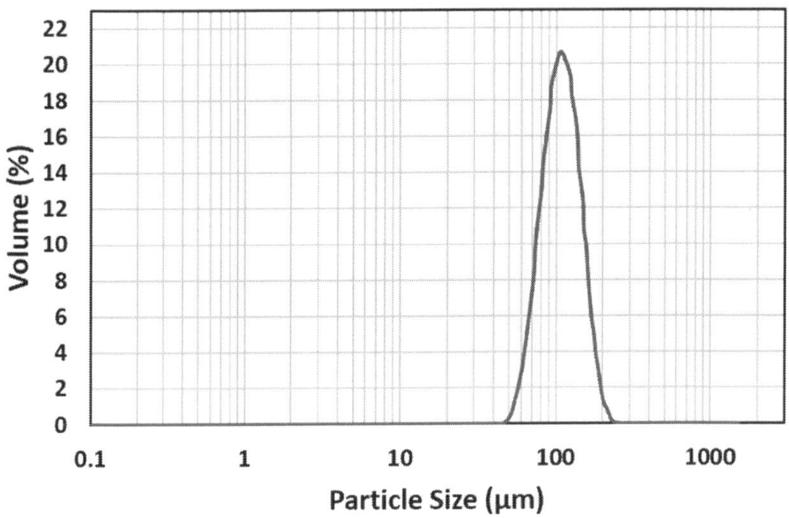

Figure 1. Powder particle size distribution measured by laser diffraction technique.

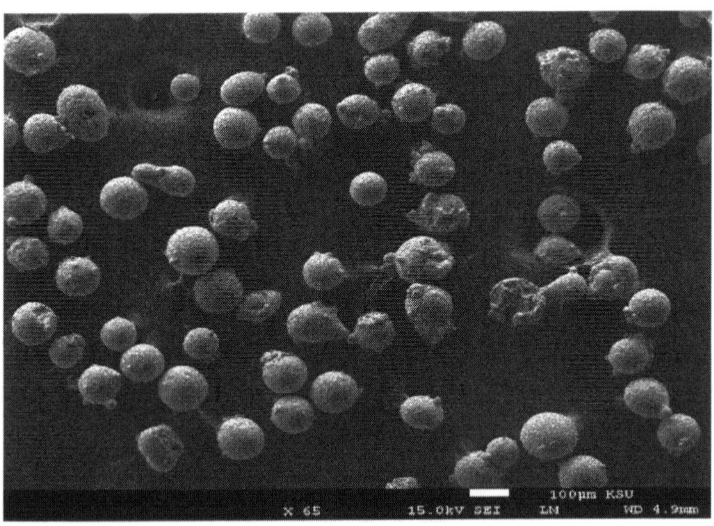

Figure 2. Scanning electron microscopy image of the powder particles before consolidation.

After the build was completed, the parts were enveloped by the surrounding powder. Because the preheating step envelops the entire area of the substrate plate, the powder around each sample gets slightly sintered. To remove this powder, the build was taken to powder recovery system and blasted with compressed air. To make the blasting more effective, loose powder from the same feedstock was fed along with the air stream.

In EBM the layer thickness was set to 90 microns. Beam current was 25 mA and voltage was set to 60 kV. The beam scan speed was 2400 mm/s. The base plate was heated up to 1050 °C before the first layer of powder was spread on it. Thereafter, the nominal build temperature was set to 1100 °C. The EBM controller maintained the vacuum to about 2 × 10−3 mBar, while the building was continued. The build was surrounded by small amount of helium (He). Ionization of He gave stability of the electron beam and at the same time held down the powder particles thus avoided powder clouding. Powder clouding known as "smoke" can result when electron charge builds up too fast on the surface of the powder particles in the bed. Figure 3 shows SEM image of a surface of the EBM produced γ-TiAl alloy. The corrosion studies were carried out on this same surface. SEM investigations did not reveal any pores of substantial size, suggesting that the consolidation of the powder has been fair enough in the EBM process. The microstructure consisted of laths of γ and α2 phases. The microstructure was uniform throughout without any clear demarcation of melt tracks. This implies that the adjacent electron beam scans overlapped at an optimum distance and that the subsequent powder layers have fused well.

2.2. Chemicals, Materials, and Electrochemical Cell

The electrochemical experiments were performed in a 3.5% NaCl solution with conventional three-electrode cell system in which

Ag/AgCl acted as a reference electrode (RE), the platinum foil as a counter electrode (CE) and the produced Ti-Aluminide alloy as the working electrode (WE). The preparation of WE involved a series of steps wherein; a blind hole of about 0.5 mm in depth and diameter was drilled on one surface of the alloy. A long copper wire with similar diameter was placed in the drilled hole and later soldered to form a rigid joint. The entire assembly was then mounted in an epoxy resin and allowed to cure for 24 h at room temperature. After curing, the other surface was polished using sandpaper of various grit sizes ranging from 180 grit to 1000 grit. The surface was later fine polished using colloidal silica solution to produce a scratch free mirror surface.

Figure 3. Microstructure of the surface of the fabricated γ-TiAl alloy.

2.3. Potentiodynamic Polarization and Electrochemical Impedance (EIS)

The electrochemical experiments were performed using Autolab system manufactured by Metrohm (PGSTAT20, Amsterdam, The Netherlands). The open circuit potential (versus Ag/AgCl) was carefully observed after immersing the working electrode in the test solution until the potential stabilized within ±1 mV. Once the potential stability was established, an EIS test was initiated and recorded. The scanning frequency was selected in the range of 100 kHz to 100 MHz, along with an ac wave of ±5 mV peak-to-peak overlaid on a dc bias potential. An equivalent circuit of Nyquist plots was deduced by fit and simulation method. The potentiodynamic polarization curves were obtained by scanning the potential in the forward direction from −0.8 to 0.0 V against Ag/AgCl at a scan rate of 1.67 mV/s. For each experimental condition, three measurements were performed to ensure the reliability and reproducibility of the data.

2.4. Surface Characterization

Scanning electron microscope (SEM) and energy dispersive X-ray (EDX) analyzer both from JEOL (Tokyo, Japan) were employed. The SEM images and the corresponding EDX profile analyses were obtained before and after immersing the alloy in the test chloride solution.

3. RESULTS AND DISCUSSION

3.1. Potentiodynamic Polarization Measurements

The corrosion behavior after different periods of exposure in an aerated 3.5% NaCl solution using potentiodynamic polarization measurement of the γ-TiAl alloy, which was produced by electron beam melting (EBM) process, was reported. Figure 4 shows the potentiodynamic polarization curves obtained for the γ-TiAl alloy after (1) 0.0 h; (2) 1.0 h; (3) 4.0 h; and (4) 24 h immersion in 3.5% NaCl solution at room temperature. The corrosion parameters obtained from the polarization curves were calculated and recorded in Table 1. The values of cathodic Tafel (βc) and anodic Tafel (βa) slopes, corrosion potential (ECorr), corrosion current density (jCorr) and polarization resistance (Rp) that are listed in Table 1 were obtained as previously reported in our earlier studies [12,13,14].

Table 1. Potentiodynamic polarization parameters obtained for γ-TiAl alloy in 3.5% NaCl solutions at room temperature.

Exposure Period/h	Parameter				
	β_a/mV·dec^{-1}	β_c/mV·dec^{-1}	E_{Corr}/mV	j_{Corr}/μA·cm^{-2}	R_p/kΩ·cm^2
0	65.66	72.81	−497	0.157	255.9
1	62.21	74.95	−538	0.102	339.7
4	60.83	70.75	−591	0.076	368.3
24	65.02	74.34	−595	0.074	440.8

It is clearly seen from Figure 4 that the current decreases in the cathodic side reaching the value of jCorr, which is the minimum value of current recorded by the alloy before the current increases again in the anodic branch under the effect of the increase in the applied potential and the aggressiveness influence of chloride ions. It is also seen that the alloy immersed for 0.0 h (curve 1) showed the highest cathodic current, jCorr and the less negative value of ECorr.

Increasing the immersion time to 1.0 h (curve 2) decreased the values of cathodic and anodic current and jCorr as well as shifted the value of ECorr towards the more negative direction. Further increases of immersion time to 4.0 h and 24 h decreases the corrosion of the alloy through decreasing its corrosion parameters. This is because the increase of immersion time decreases the uniform attack of γ-TiAl alloy in the chloride solution. This was further confirmed by the parameters that are depicted in Table 1, where the increase of immersion time shifted ECorr to the more negative value and decreased the value of jCorr as well as increased the corrosion resistance values, Rp. The high Rp values correspond to the working electrode opposing any attempt to modify its equilibrium state, resulting in a low rate of titanium ion release and oxide growth. This implies that this alloy has very good corrosion resistance. This behavior was clearly noted for γ-TiAl in Ringer's solution and is in agreement with the results found with potentiodynamic polarization [10].

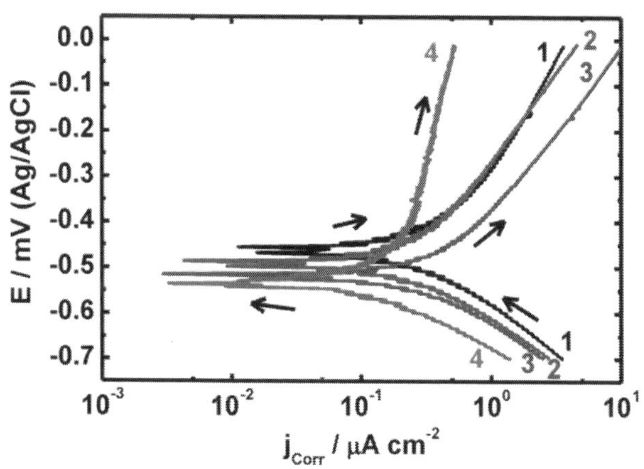

Figure 4. Potentiodynamic polarization curves of γ-TiAl alloy after (1) 0.0 h; (2) 1.0 h; (3) 4.0 h and (4) 24 h immersion in 3.5% NaCl solution at room temperature.

In order to report the effect of raising temperature on the corrosion of γ-TiAl alloy in the chloride test solution, 3.5% NaCl, potentiodynamic polarization measurements were also carried out. The polarization curves obtained for γ-TiAl alloy in 3.5% NaCl solution at (1) 30 °C; (2) 40 °C; (3) 50 °C; and (4) 60 °C, respectively are displayed in Figure 5. The values of the corrosion parameters, namely βc, βa, ECorr, jCorr, and Rp obtained from the curves shown in Figure 5 are listed in Table 2. The increase of temperature was found to increase the corrosion of the alloy via increasing the cathodic currents, anodic currents and jCorr, while decreased the values of Rp and shifted ECorr to the less negative direction. The corrosion parameters shown in Table 2 proved also that the increase of temperature significantly increases the corrosion of the alloy in the chloride test solution. This is because the increase of temperature increases the activation of the surface of the alloy, which reflects on its fast dissolution and thus the increase of its corrosion [15,16].

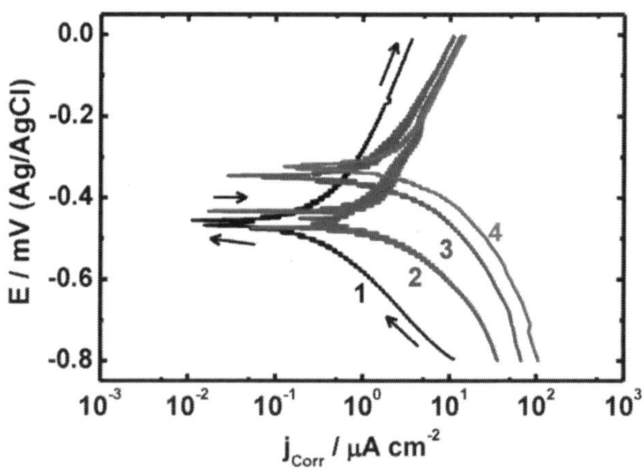

Figure 5. Potentiodynamic polarization behavior for γ-TiAl alloy in 3.5% NaCl solution at (1) 30 °C; (2) 40 °C; (3) 50 °C; and (4) 60 °C, respectively.

Table 2. Potentiodynamic polarization parameters obtained for γ-TiAl alloy in 3.5% NaCl solutions after different temperatures.

Temperature/°C	Parameter				
	β_a/mV·dec^{-1}	β_c/mV·dec^{-1}	E_{Corr}/mV	j_{Corr}/μA·cm^{-2}	R_p/kΩ·cm^2
20	65.66	72.81	−497	0.10	255.9
30	66.91	74.74	−438	0.25	166.7
40	60.78	74.09	−356	0.35	105.8
50	61.33	70.82	−330	0.44	70.88
60	64.37	73.85	−322	0.73	50.93

3.2. Electrochemical Impedance Spectroscopy (EIS)

Electrochemical impedance spectroscopy (EIS) method has been normally used to study the corrosion behavior and passivation phenomena for different metals and alloys in a variety of corrosive environments [17,18,19,20,21,22]. The corrosion behavior of the γ-TiAl alloy samples were carried out in freely aerated stagnant solution of 3.5% NaCl after various immersion periods of time at room temperature and the Nyquist plots obtained at the open-circuit potential are shown in Figure 6. Similar plots were also obtained for the alloy at different temperatures of 20, 30, 40, 50, and 60 °C and spectra are shown in Figure 7. The Nyquist plots shown in Figure 6 and Figure 7 were analyzed by fitting the experimental results to an equivalent circuit model shown in Figure 8.

The elements of the circuit shown in Figure 8 are; the solution resistance (RS), the constant phase elements (Q) with their n value close to 1.0 refers to a double layer capacitance at the electrode/solution interface (Cdl), the polarization resistance (Rp1) that can also be defined as the charge transfer resistance at the electrode/solution interface, the double layer capacitance (Cdl), and another polarization resistance (Rp2) that refers to the resistance of the interface between the corrosion products and the solution. The values of these parameters were determined at room temperature after different immersion periods of time and listed in Table 3. While, the values of the impedance parameters for the γ-TiAl alloy, which were obtained at different temperatures are depicted in Table 4. Among these parameters, Rp1 and Rp2 are the factors that determine the corrosion resistance of alloys. This is because the values of Rp1 and Rp2 are inversely proportional to jCorr, and hence, high values of Rp correspond to low corrosion rates.

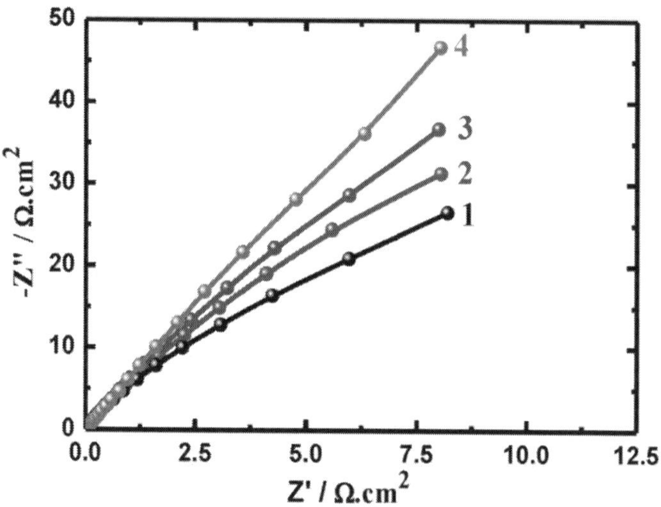

Figure 6. Nyquist plots for γ-TiAl alloy after (1) 0.0 h; (2) 1.0 h; (3) 4.0 h; and (4) 24 h immersion in 3.5% NaCl solution at room temperature.

The shapes of Nyquist plots shown in Figure 6 are similar at each exposure period with one depressed semicircle indicating that the corrosion mechanism of the alloy is similar for all regions. It is observed that the diameter of the semicircle remarkably increased with the increase in immersion time, which indicates that the corrosion resistance of the alloy increases with time. The values of Rp listed in Table 3 also increases with the increase of exposure period of time and reveal that the corrosion of the alloy decreases with increasing immersion time. It is also seen that the diameter of the arc, giving the charge-transfer resistance of corrosion reaction, exceeds 100 Ωcm2. This high Rp value is most probably due to the formation of a stable oxide layer on the γ-TiAl alloy and that has an excellent corrosion resistance in the 3.5 wt. % NaCl solution. It is worth mentioning also that there is a difference between the values of Rp1 obtained here than those obtained by polarization (Table 1) and this is because polarization is an activation technique, while EIS data are taken at free potential. Moreover, the decrease of the value of the film capacitance (Cdl) with increasing immersion time is resulted in turn by the stability of the passive oxide layer formed on the titanium aluminide alloy. Lower values of Cdl, especially at long immersion periods of time (Table 3) are corresponding to a slow growth of the oxide film, indicating long-term stability of the passive layer [23,24]. Thus the results from EIS measurements are in good agreement with the obtained polarization data for the γ-TiAl alloy after its immersion for different periods of time in 3.5% NaCl solution.

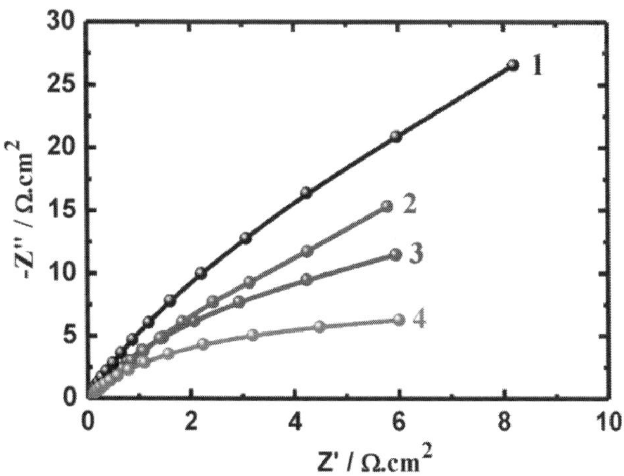

Figure 7. Nyquist plots for γ-TiAl alloy in 3.5% NaCl solution at (1) 30 °C; (2) 40 °C; (3) 50 °C; and (4) 60 °C, respectively.

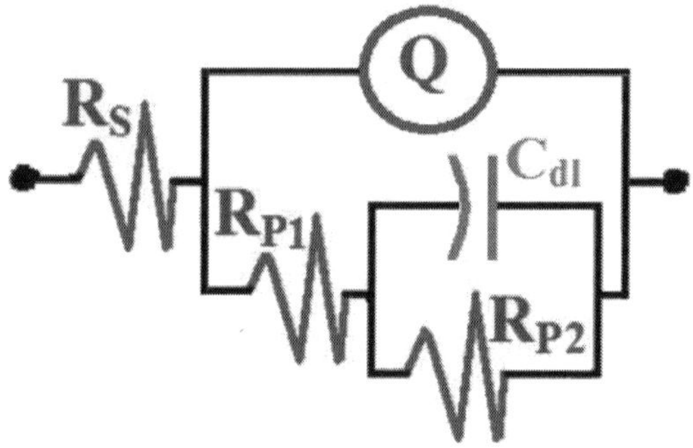

Figure 8. Equivalent circuit model that fits the EIS experimental data.

Table 3. EIS parameters obtained for γ-TiAl alloy in 3.5% NaCl solutions at room temperature.

Exposure Period/h	$R_S/\Omega\cdot cm^2$	Q		$R_{P1}/\Omega\cdot cm^2$	$C_{dl}/F\cdot cm^{-2}$	$R_{P2}/\Omega\cdot cm^2$
		$Y_Q/F\cdot cm^{-2}$	n			
0	9.22	0.3251	0.73	4.83	0.2992	15.82
1	9.83	0.2978	0.86	6.44	0.2814	21.13
4	9.99	0.2641	0.84	14.2	0.2656	25.35
24	10.03	0.2489	0.89	19.7	0.2115	37.57

Table 4. EIS parameters obtained for γ-TiAl alloy in 3.5% NaCl solutions after different temperatures.

Temperature/°C	$R_S/\Omega\cdot cm^2$	Q		$R_{P1}/\Omega\cdot cm^2$	$C_{dl}/F\cdot cm^{-2}$	$R_{P2}/\Omega\cdot cm^2$
		$Y_Q/F\cdot cm^{-2}$	n			
20	9.22	0.325	0.73	4.83	0.2992	15.82
30	8.99	0.338	0.83	4.36	0.3199	12.13
40	8.42	0.345	0.85	3.50	0.3537	9.38
50	8.35	0.383	0.87	2.99	0.3735	7.59
60	8.12	0.394	0.89	2.06	0.4849	5.94

It is clearly seen from Figure 7 for the γ-TiAl alloy at different temperatures that there is only distorted semicircle at all temperatures, whose diameter gets smaller with increasing the temperature. This indicates that the corrosion of the alloy increases with increasing temperature. This was also confirmed by the values of the EIS parameters shown in Table 4. Where the values of Rp1 and Rp2 decreased significantly with temperature and proves that the increase of temperature increases the corrosion for the alloy in the chloride test solution. The value of the constant phase elements (CPEs, Q) with their n values among 0.73 and 0.89 at all temperatures indicate that the CPE represents a double layer capacitor (Cdl). This means that the dissolution of the alloy surface is limited by the mass transport, which increases with increasing the solution temperature as was indicated by the increase of Cdl value with increasing the temperature. The increased corrosion with temperature can be explained on the basis that the increase of temperature decreases the opportunity for the oxide film formation and thus decreases the corrosion resistance of the surface of γ-TiAl alloy.

The Bode (a) impedance of the interface and (b) the phase angle plots for γ-TiAl alloy after (1) 0.0 h; (2) 1.0 h; (3) 4.0 h; and (4) 24 h immersion in 3.5% NaCl solution at room temperature are shown in Figure 9. It is seen from Figure 9a that the impedance of the interface (|Z|) for the alloy shows lower values, especially at high frequency, increases with decreasing frequency, and the highest values of |Z| are recorded at the lowest frequency values. Increasing the immersion time is obvious to increase the values of |Z|, which indicates that the increase of immersion time increases the corrosion resistance of the alloy in the chloride solution. This was further confirmed by plotting the change of the degree of phase angle (Φ) with frequency as shown in Figure 9b. It is noted from Figure 9b that the increase of immersion time of the alloy before measurements increases the maximum degree of Φ and that in turn confirms the increase of the passivation of the surface of the aluminum alloy with the increase of time of exposure. However, there is a general trend of the impedance modulus that increases with increasing time of immersion and decreases with increasing temperature, as was also provided by the potentiodynamic results. The EIS data thus confirm the results obtained from the polarization measurements that the increase of immersion time decreases the corrosion of the investigated alloy and that the increase of temperature increases its corrosion.

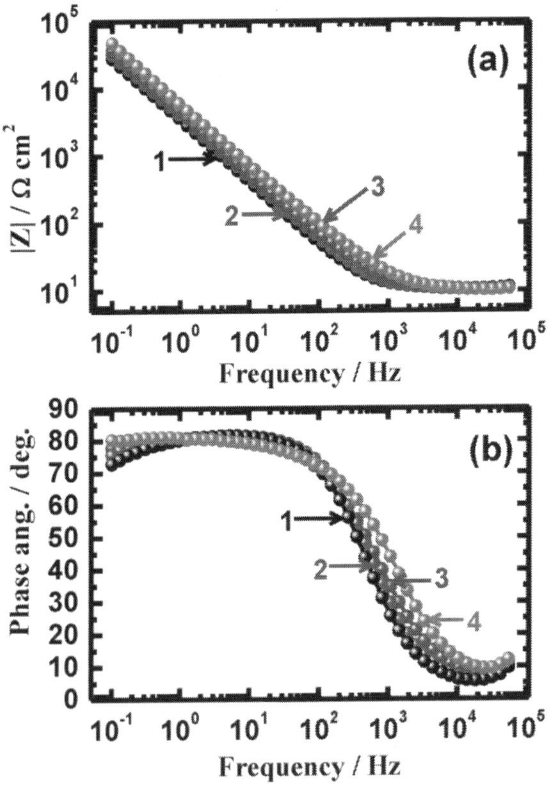

Figure 9. Bode (a) impedance of the interface and (b) the phase angle plots for γ-TiAl alloy after (1) 0.0 h; (2) 1.0 h; (3) 4.0 h; and (4) 24 h immersion in 3.5% NaCl solution at room temperature.

3.3. Scanning Electron Microscopy (SEM) and Energy Dispersive X-ray (EDX) Analyses

SEM and EDX investigations were carried out on the surface of the γ-TiAl alloy before and after its immersion in the chloride test solution, 3.5% NaCl, in order to report the effect of the chloride solution on the surface of the alloy and to quantify the alloying elements of the alloy before and after corrosion. Figure 10 shows the SEM micrograph and EDX profile analysis for the γ-TiAl alloy before its immersion in 3.5% NaCl solution. The SEM image shows that the surface is clear and homogenous. The EDX corresponding to the SEM image depicted in Figure 10 indicates that the atomic percents for the elements found on the surface of the γ-TiAl alloy were as follows: 46.08% Al, 49.63% Ti, 2.35% Nb, and 1.94% Cr.

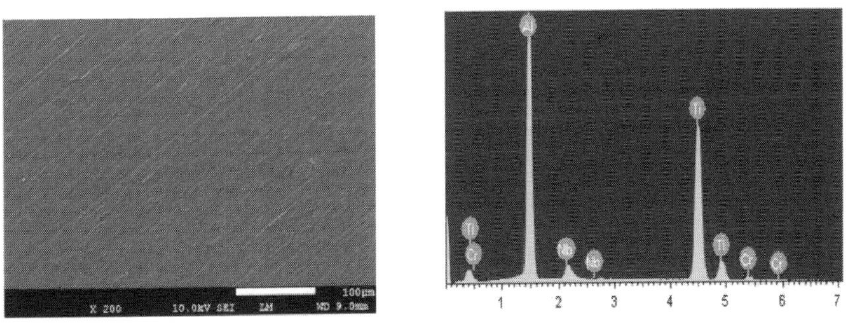

Figure 10. SEM micrograph and EDX profile analysis for the γ-TiAl alloy before its immersion in 3.5% NaCl solution.

The SEM and EDX profile analysis taken after immersing the γ-TiAl alloy for an hour in the chloride solution at room temperature followed by scanning the potential in the forward direction from −0.08 to 0.0 V (Ag/AgCl) are shown in Figure 11. The surface of the alloy is shown to be affected by the severity attack of the chloride ions present in the solution and also the increase of the applied potential towards the positive direction. Here, the SEM image (Figure 11) indicates that the surface is different from the surface shown in Figure 10 from the sample before immersion. Several shallow pits have resulted from immersing the γ-TiAl alloy in 3.5% NaCl solution and from sweeping the potential. The EDX corresponding to the SEM image depicted in Figure 10 indicates that the atomic percents for the elements found on the surface of the γ-TiAl alloy were as following 40.18% Al, 46.69% Ti, 2.25% Nb, 1.65% Cr, and 9.22% O. It is obviously noted from this EDX data that the percent of Al decreased compared to its percent before immersion. This means the corrosion of the alloy might have resulted from the dissolution of Al under the chloride ions attack and the increase of the applied anodic potential. Also, the presence of oxygen is most probably resulting from the formation of aluminum oxide layer on the surface of the alloy, which in turn leads to decreasing the corrosion of the γ-TiAl alloy in 3.5% NaCl solution. The atomic percent for Nb and Cr were almost unchanged before and after immersion, which confirms also that Al is responsible for the corrosion and protection of the alloy.

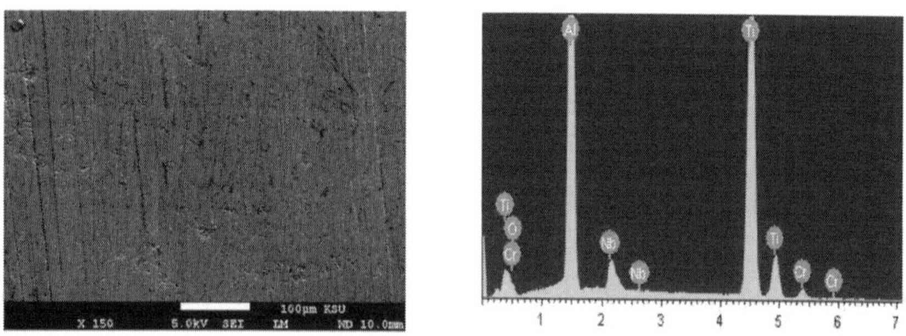

Figure 11. SEM micrograph and EDX profile analysis for the γ-TiAl alloy after its immersion 1 h in 3.5% NaCl solution followed by sweeping the potential from −0.8 V to 0.0 V vs. Ag/AgCl.

4. CONCLUSIONS

γ-TiAl alloy was produced from metallic powders and processed using electron beam melting (EBM) technique. The corrosion behavior of the produced γ-TiAl alloy was investigated at different temperatures and after different exposure periods of time in 3.5% NaCl solutions. The corrosion tests were carried out using potentiodynamic polarization and EIS techniques. The surface of γ-TiAl alloy was investigated before and after exposure to the chloride solution by using SEM and EDX analyses. The results show that the EBM produced γ-TiAl alloy has excellent corrosion resistance confirmed by the high values of polarization resistance and corrosion potential and low values of corrosion current and corrosion rate. It has been found that the increase in immersion time increases the corrosion resistance of the alloy via decreasing the anodic, cathodic, and corrosion currents and corrosion rate, as well as shifts the corrosion potential towards the positive values. On the other hand, the increase of solution temperature from 20 to 60 °C as was indicated by the increase of corrosion parameters is measured by polarization and impedance techniques. SEM/EDX investigations confirmed that the formation of an oxide film on the surface of the alloy is responsible for the corrosion protection of the γ-TiAl alloy, while its corrosion results from the dissolution of aluminum; furthermore, the percentages of Nb and Cr stay almost unchanged before and after the corrosion of the alloy.

ACKNOWLEDGMENTS

This project was funded by the National Plan for Science, Technology, and Innovation (MAARIFAH), King Abdulaziz City for Science and Technology, Kingdom of Saudi Arabia, Award Number (11-ADV1494-02).

AUTHOR CONTRIBUTIONS

A.H. Seikh designed the work and conducted the corrosion experiments. A. Mohammed fabricated the γ-TiAl alloy by EBM and helped in writing the draft of manuscript. E.-S.M. Sherif participated in performing the corrosion tests and wrote the draft manuscript and edited the final manuscript. A. Al-Ahmari helped in manufacturing the alloy and writing the manuscript.

REFERENCES

1. Murr, L.E.; Gaytan, S.M.; Ceylan, A.; Martinez, E.; Martinez, J.L.; Hernandez, D.H.; Machado, B.I.; Ramirez, D.A.; Medina, F.; Collins, S.; et al. Characterization of titanium aluminide alloy components fabricated by additive manufacturing using electron beam melting. Acta Mater. 2010, 58, 1887–1894.

2. Donachie, M.J., Jr. Titanium—A Technical Guide; ASM International: Geauga County, OH, USA, 2000.

3. Wu, X. Review of alloy and process development of TiAl alloys. Intermetallics 2006, 14, 1114–1122.

4. Draper, L.S.; Das, G.; Locci, I.; Whittenberger, J.D.; Lerch, B.A.; Kestler, H. Gamma Titanium Alumindes; Kim, Y.-W., Clemens, H., Rosenberger, A.H., Eds.; TMS: Warrendale, PA, USA, 2003; pp. 207–212.

5. Liu, X.; Chu, P.; Ding, C. Surface modification of titanium, titanium alloys, and related materials for biomedical applications. Mater. Sci. Eng. R 2004, 47, 49–121.

6. Tetsui, T.; Shindo, K.; Kaji, S.; Kobayashi, S.; Takeyama, M. Fabrication of TiAl components by means of hot forging and machining. Intermetallics 2005, 13, 971–978.

7. Tetsui, T.; Shindo, K.; Kaji, S.; Kobayashi, S.; Takeyama, M. Strengthening a high-strength TiAl alloy by hot-forging. Intermetallics 2003, 11, 299–306.

8. Wegmann, G.; Gerling, R.; Schimansky, F.P.; Clemens, H.; Bartels, A. High-temperature mechanical properties of hot isostatically pressed and forged gamma titanium aluminide alloy powder. Intermetallics 2002, 10, 511–517.

9. Kuang, J.P.; Harding, R.A.; Campbell, J. Microstructures and properties of investment castings of γ-titanium aluminide. Mater. Sci. Eng. A 2002, 329–331, 31–37.

10. Delgado-Alvarado, C.; Sundaram, P.A. Corrosion evaluation of Ti-48Al-2Cr-2Nb (at. %) in Ringer's solution. Acta Biomater. 2006, 2, 701–708.

11. Gurrapa, I. Degradation of Ti-24Al-15Nb alloys under different environmental conditions. Intermetallics 2003, 11, 867–871.

12. Sherif, E.-S.M.; Almajid, A.A.; Latif, F.H.; Junaedi, H. Effects of Graphite on the Corrosion Behavior of Aluminum-Graphite Composite in Sodium Chloride Solutions. Int. J. Electrochem. Sci. 2011, 6, 1085–1099.

13. Latief, F.H.; Sherif, E.-S.M.; Almajid, A.A.; Junaedi, H. Fabrication of exfoliated graphite nanoplatelets-reinforced aluminum composites and evaluating their mechanical properties and corrosion behavior. J. Anal. Appl. Pyrolysis 2011, 92, 485–492.

14. Sherif, E.-S.M. Corrosion and Corrosion Inhibition of Aluminum in Arabian Gulf Seawater and Sodium Chloride Solutions by 3-

Amino-5-Mercapto-1,2,4-Triazole. Int. J. Electrochem. Sci. 2011, 6, 1479–1492.

15. Latief, F.H.; Sherif, E.-S.M.; Kakehi, K. Role of Aluminide Coating on Oxidation Resistance of Ni-Based Single Crystal Superalloy at 900 °C. Int. J. Electrochem. Sci. 2015, 10, 1873–1882.

16. Zubaidy, E.A.H.A.; Mohammad, F.S.; Bassioni, G. Effect of pH, Salinity and Temperature on Aluminum Cookware Leaching during Food Preparation. Int. J. Electrochem. Sci. 2011, 6, 6424–6441.

17. Darowicki, K.; Krakowiak, S.; Ślepski, P. Evaluation of pitting corrosion by means of dynamic electrochemical impedance spectroscopy. Electrochim. Acta 2004, 49, 2909–2918.

18. Floyd, F.L.; Avudaiappan, S.; Gibson, J.; Mehta, B.; Smith, P.; Provder, T.; Escarsega, J. Using electrochemical impedance spectroscopy to predict the corrosion resistance of unexposed coated metal panels. Prog. Org. Coatings 2009, 66, 8–34.

19. Sherif, E.-S.M. Electrochemical investigations on the corrosion inhibition of aluminum by 3-amino-1,2,4-triazole-5-thiol in naturally aerated stagnant seawater. J. Ind. Eng. Chem. 2013, 19, 1884–1889.

20. Sherif, E.-S.M.; Abdo, H.S.; Khalil, K.A.; Nabawy, A.M. Corrosion Properties in Sodium Chloride Solutions of Al-TiC Composites in situ Synthesized by HFIHF. Metals 2015, 5, 1799–1811.

21. Shinde, V.; Patil, P.P. Evaluation of corrosion protection performance of poly (o-ethyl aniline) coated copper by electrochemical impedance spectroscopy. Mater. Sci. Eng. B 2010, 168, 142–150.

22. Gao, W.; Cao, S.; Yang, Y.; Wang, H.; Li, J.; Jiang, Y. Electrochemical impedance spectroscopy investigation on indium tin oxide films

under cathodic polarization in NaOH solution. Thin Solid Films 2012, 520, 6916–6921.

23. Mansfeld, F.; Fernandes, J.C.S. Impedance spectra for aluminum 7075 during the early stages of immersion in sodium chloride. Corros. Sci. 1993, 34, 2105–2108.

24. Mansfeld, F.; Han, L.T.; Lee, C.C.; Chen, C.; Zhang, G.; Xiao, H. Analysis of electrochemical impedance and noise data for polymer coated metals. Corros. Sci. 1997, 39, 255–279.

CHAPTER 9

Enhancing Corrosion and Wear Resistance of AA6061 by Friction Stir Processing with Fe78Si9B13 Glass Particles

Lingyu Guo [1], Yan Liu [1], Kechang Shen [1], Chaoqun Song [1], Min Yang [1], Kibuem Kim [2] and Weimin Wang [1,*]

[1] Key Laboratory for Liquid-Solid Structural Evolution and Processing of Materials, Ministry of Education, Shandong University, Jinan 250061, China

[2] Hybrid Materials Center (HMC), Faculty of Nanotechnology and Advanced Materials Engineering, Sejong University, 209 Neungdong-ro, Gwangjin-gu, Seoul 143-747, Korea

ABSTRACT

The AA6061-T6 aluminum alloy samples including annealed Fe78Si9B13 particles were prepared by friction stir processing (FSP) and investigated by various techniques. The Fe78Si9B13-reinforced particles are uniformly dispersed in the aluminum alloy matrix. The XRD results indicated that the lattice parameter of α-Al increases and the preferred orientation factors F of (200) plane of α-Al reduces after friction stir processing. The coefficient of thermal expansion (CTE) for FSP samples increases at first with the temperature but then decreases as the temperature further increased, which can be explained by the dissolving of Mg and Si from β phase and Fe78Si9B13 particles. The corrosion and wear resistance of FSP samples have been improved compared with that of base metal, which can be attributed to the reduction of grain size and the CTE mismatch between the base metal and reinforced particles by FSP, and the lubrication effect of Fe78Si9B13 particles also plays a role in improving wear resistance. In particular, the FSP sample with reinforced particles in amorphous state exhibited superior corrosion and wear resistance due to the unique metastable structure.

KEYWORDS

aluminum alloy; Fe78Si9B13; friction stir processing; corrosion; wear

1. INTRODUCTION

Aluminum and its alloys are used extensively in numerous fields due to their low densities and high strength to weight ratio [1,2]. Nevertheless, their poor resistance to wear and corrosion causes certain limitations for their application [3]. Metal matrix composites are a new type of material that exhibit good wear and corrosion resistance properties as compared to the matrix [4,5].

As wear and corrosion are surface-dependent degradation, they can be improved by a suitable modification of surface microstructure and/or composition. A proper technique can be employed to refine the microstructure and homogeneous dispersion of reinforcements on metallic surface. With conventional surface modification techniques it is difficult to improve the dispersion of reinforcement particles on the metal surface [6]. According to previous research

[7,8], thermal spraying and laser beam techniques were used to prepare surface composites, but the creation of an unfavorable phase contributed to the degeneration of the properties. These techniques were operated at high temperature and it was difficult to avoid a reaction between the reinforcements and the matrix, which leads to the formation of an adverse phase. Thus, a fabrication process below the melting temperature of the matrix is critical to avoid the complications mentioned above.

Recently, friction stir processing (FSP), a successful surfaced modification technique derived based on the principles of friction stir welding (FSW) by Mishra et al. [9], has attracted much attention [10,11]. During the processing, a non-consumable rotating tool is plunged into the interface between two plates to be joined and traversed along the line of the joint with a specially designed pin and shoulder. The severe plastic deformation and material flow in the stirred zone can be utilized to achieve bulk alloy modification via mixing of other elements or the second phase into the stirred alloys, accompanied with significant grain refinement [9]. Friction stir processing is a solid state processing technique to obtain surface composites. In previous studies [12,13,14,15], the most widely used reinforcements are ceramics, such as Al2O3 or SiC in the conventional metal matrix composites. There is little research about using amorphous alloy as a reinforced material. The amorphous materials have superior properties such as high strength, high hardness, and excellent corrosion resistance due to their disordered and metastable structure [16,17,18,19].

In the present study, we focus on the development of a new AA6061-T6 alloy matrix reinforced with Fe78Si9B13 particles, which are fabricated by a friction stir process. The microstructure, corrosion resistance, wear behavior, and magnetic properties of the composites and base metal are investigated by different techniques.

2. EXPERIMENTAL PROCEDURE

In this study, the base metal employed is a 4 mm thick aluminum alloy AA6061-T6, which has a chemical composition of (in wt.%) Mg: 0.95; Si: 0.54; Fe: 0.22; Mn: 0.13; Cu: 0.17; Cr: 0.09; Zn: 0.08; Ni: 0.02; Ti: 0.01; Al: balance [20]. Using manual drilling machines we created a series of blind holes in the aluminum alloy surface, with a diameter and depth of 3 mm. The reinforced particles are Fe78Si9B13 metal glasses after annealing and high energy milling processes. The annealing temperatures are 300 °C and 700 °C, respectively. The milling parameters are selected as following: the ratio of ball to powder is 10:1, frequency is 20 Hz, and the milling temperature is −196 °C. The FSP machine was used for fabricating in situ hybrid surface composites. The parameters for FSP include a traveling speed (v) of 45 mm/min, a rotational rate (w) of 900 r/min, a descending distance of 3 mm, and four stirrings. The base metal and the composites reinforced with particles annealed at 300 °C and 700 °C were named S000, S300, and S700, respectively.

The phase components of samples were investigated by X-ray diffractometer with Cu target (D/MAX 2500/PC, Rigaku, Tokyo, Japan, Cu Kα, λ = 0.154056 nm) and Co target (X'Pert PRO, PANalytical, Almelo, the Netherlands, Co Kα, λ = 0.178897 nm). The lattice constant is calculated by the extended Bragg equation [21]:

$$a_0 = \frac{\lambda}{2\sin\theta}\sqrt{h^2+k^2+l^2}, \qquad (1)$$

where λ is the radiation wavelength; θ is the diffraction angle; and h, k,and l are the crystal plane indices. In present samples, the preferred orientation factors F of (200) plane of α-Al are investigated by the Lotgering method [22]:

$$F = \frac{P - P_0}{1 - P_0}, P = \frac{\sum I_{(h00)}}{\sum I_{(hkl)}} \text{ and } P_0 = \frac{\sum I_{0(h00)}}{\sum I_{0(hkl)}}, \tag{2}$$

where I(h00) and I(hkl) are the integral intensities of diffraction peaks for composites samples and randomly oriented samples, respectively; and P and P0 are the ratios of integral intensities of (h00) planes to those of all (hkl) planes for composites samples and randomly oriented samples, respectively.

The microstructure of samples was examined by scanning electron microscopy (SEM, SUPRA55, Zeiss, Germany). Thermal expansion curves were measured by a dilatometer (DIL, Netzsch 402C, Netzsch, Shanghai, China) with a constant heating rate 20 K/min, performed under flowing high purity argon to protect the samples from oxidation.

The electrochemical measurements were carried out by CHI660E advanced electrochemical workstation with a scanning rate of 1 mV/s at room temperature. There was a typical three-electrode system in electrochemical measurements: working electrode, platinum counter electrode and Hg|Hg2Cl2 (SCE) reference electrode. The electrolyte is 3.5 wt% NaCl solutions.

The wear behavior of the base metal and hybrid surface composites was measured using a wear apparatus (M-2000, Yinuo Testing Machine, Jinan, China) at room temperature. The wear test was conducted at a rotation rate of abrasive wheel of 200 r/min, normal force of 120 N, and sliding time of 10 min.

3. RESULTS

Figure 1a illustrates an overview of the side of the FSP plate in contact with the shoulder tool. This surface shows semicircular features, similar to those induced by a conventional milling process without cracking and other disfigure. The XRD patterns of the reinforced particles are also shown in Figure 1b, explaining that the particles annealed at 300 °C are in the amorphous state, and those annealed at 700 °C are in the crystalline state. Figure 1c,d show the morphology of the samples with the reinforced particles annealing at 300 °C and 700 °C, denoted as S300 and S700, respectively. It can be found that the reinforced particle is relatively uniform dispersed in the aluminum alloy matrix, and the size of the particle annealed at 700 °C is smaller than that of the particle annealed at 300 °C.

Figure 2 shows the XRD patterns of AA6061 FSP samples with Fe78Si9B13 particles (S300 and S700) as well as the base alloy (denoted S000). The main diffraction peaks are α-Al phases, and the other intermetallic phases are not found due to the less content. The (111), (200), (220), and (311) peaks of α-Al are easily identified, and the intensity of the (200) peak changed obviously. The inset shows that the peak of diffraction (2θ) of (200) plane of α-Al in S300 shifts to the left after FSP. For clarity, the angle of diffraction (2θ), lattice parameter a0, and preferred orientation factors F of (200) plane of α-Al are listed in Table 1. According to Equation (1), the lattice constants of the (200) plane of α-Al for the three samples are in the order S300 > S700 > S000. However, the preferred orientation factors F of the (200) plane of α-Al for the samples are in the order S300 < S700 < S000, implying that the initial rolling texture was destroyed during friction stir processing. In addition, the variation of XRD data with the Co target is confirmed by that with the Cu target.

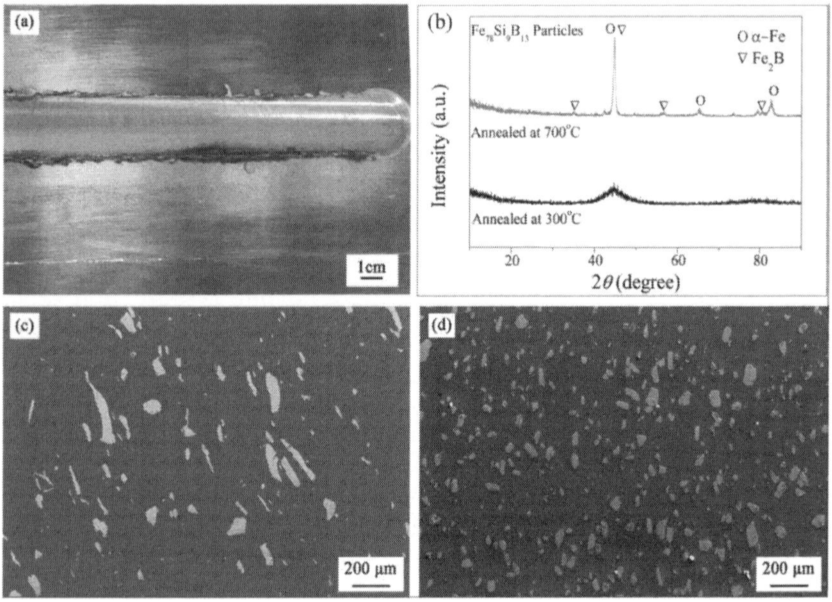

Figure 1. Topography of FSP samples: (a) Macrograph of samples after friction stir processing; (b) XRD patterns of the reinforced particles; (c) SEM photomicrograph of S300; (d) SEM photomicrograph of S700.

The thermal expansion curves and the variation of experimental coefficient of thermal expansions (CTE) obtained for base metal and FSP samples are shown in Figure 3. The three samples have an approximately linearly varying trend: the thermal expansion ratio increases with increasing temperature. In addition, the CTE of S000 increases with increasing temperature, but there is a plateau at about 300 °C. By contrast, the CTE of FSP samples at first increases with the temperature but then decreases as the temperature is further increased, and the temperature at which the CTE begins to decrease for S300 is about 350 °C, which is higher than that for S700, which is about 300 °C. Furthermore, the maximal CTEs for S300 and S700 are 2.87 × 10−5 K−1 and 3.04 × 10−5 K−1, respectively, and that of S000 at platform is 2.71 × 10−5 K−1, which is lower than the former values. Moreover, at 550 °C, the CTE of S000 is about 3.18 × 10−5 K−1, which is higher than that of S300 and S700, which are 2.48 × 10−5 K−1 and 2.71 × 10−5 K−1, respectively, i.e., in the order S300 < S700 < S000.

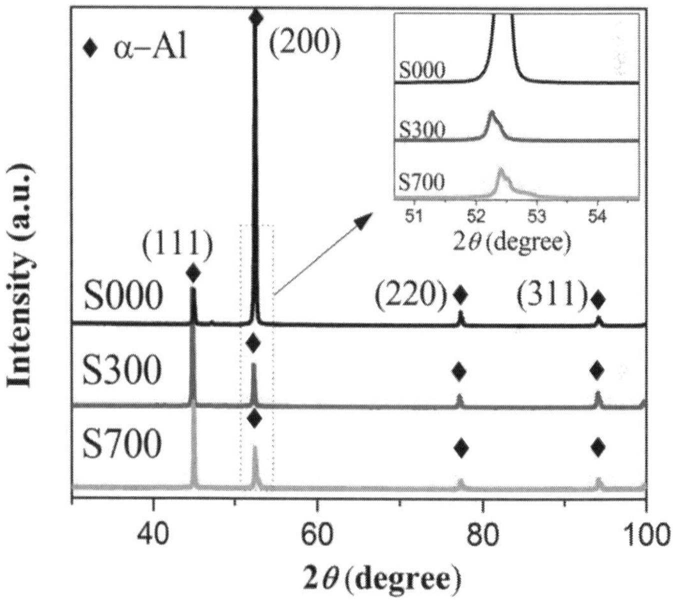

Figure 2. The XRD diffraction patterns for base metal (S000) and FSP samples (S300 and S700).

Table 1. The preferred orientation factors F, lattice parameter a0, and angle of diffraction (2θ) of the (200) plane of α-Al of samples obtained from X-ray diffraction.

Sample	2θ (degree)	a_0 (angstrom)	$(a_0 - a_0^*)/a_0^*$ (%)	F	F'
S000	52.434	4.0495	0.003	0.7530	0.7462
S300	52.252	4.0626	0.330	0.0031	0.0293
S700	52.406	4.0515	0.052	0.1023	0.1009

Notes: 2θ, a_0, and F were tested by the Co target; F' was tested by the Cu target; and a_0^* was the lattice constant of pure Al.

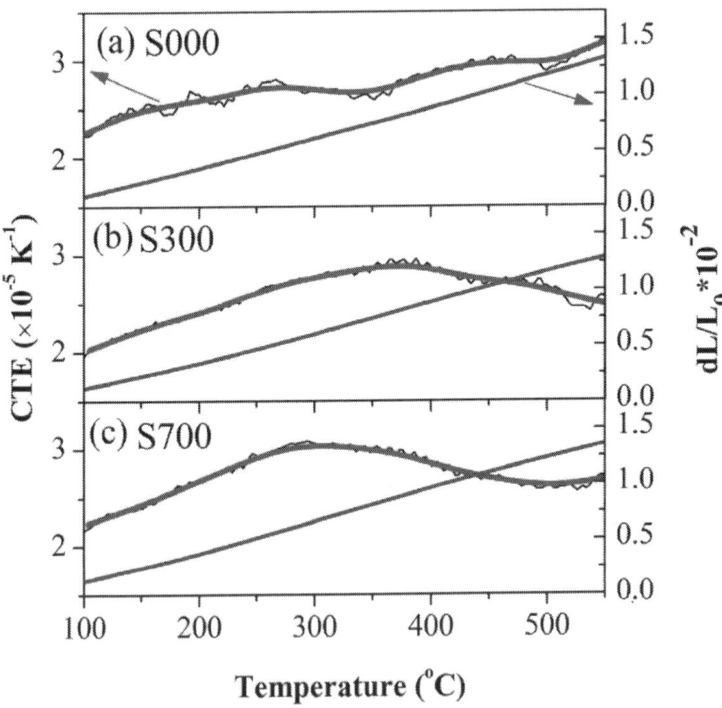

Figure 3. Thermal expansion curves (blue) and the variation of experimental coefficient of thermal expansions (CTE) (red) for (a) base metal (S000) as well as (b) S300 and (c) S700 FSP samples.

The cyclic anodic polarization curves of the base metal and FSP samples with a scanning rate of 1 mV/s, obtained in 3.5 wt% NaCl solutions, are shown in Figure 4. As can be seen from the curves, an obvious current plateau is observed in each curve, which is associated with passive film formation during the anodic polarization. The electrochemical parameters such as corrosion potential (Ecorr), corrosion current density (icorr), pitting potential (Epit), protection potential (Epp), and recorrosion current density (ircorr) are collected in Figure 5. The Ecorr of the three samples are also in the order S300 > S700 > S000, and icorr is quite the opposite, indicating that the anodic reaction of FSP samples occurs with more difficulty in the electrochemical test. Furthermore, the Epit and Epp of S300 and S700 are more positive than that of S000, and S300 has the highest Epit and Epp, which is consistent with the measured Ecorr. Here, the concept of protection potential (Epp) represents the potential at which pre-existing pits cease to propagate, that is, all the propagating pits are repassivated [23]. The Epit and Epp are related to pitting corrosion resistance, and the higher Epit and Epp indicate higher resistance to pitting corrosion. The ircorr represents the current density at which the samples incur corrosion after the repassivation region. The ircorr of FSP samples is lower than base metal, and the ircorr of S300 is the lowest, i.e., in the same order of icorr. Combined with the above measured electrochemical parameters, it is found that the corrosion resistance of the three samples is in the order S300 > S700 > S000.

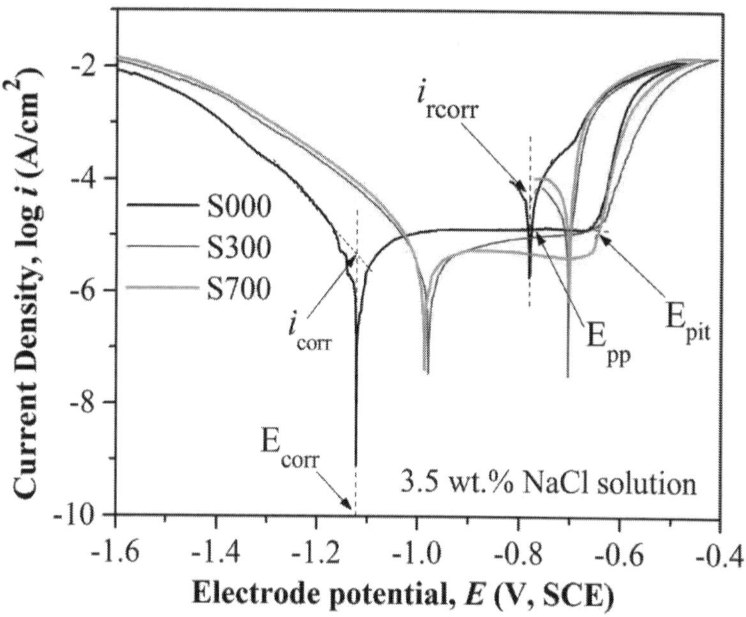

Figure 4. Cyclic anodic polarization curves of base metal (S000) and FSP samples (S300 and S700).

The SEM images of electrochemical corroded surfaces of base metal and FSP samples are shown in Figure 6. The measured samples were polarized until −0.4 VSCE with a scanning rate of 1 mV/s and without backward scanning (the polarization curves are not shown here). Many corrosion pits of 50 μm in diameter appear on the surface of S000; by contrast, there are fewer corrosion pits of smaller size (~20 μm) on the surface of S300 and S700. With further investigation, the pits on the surface of S300 are shown to be much sparser than that of S700. The pits on the base metal are deeper than on the FSP samples, accompanied by obvious cracking.

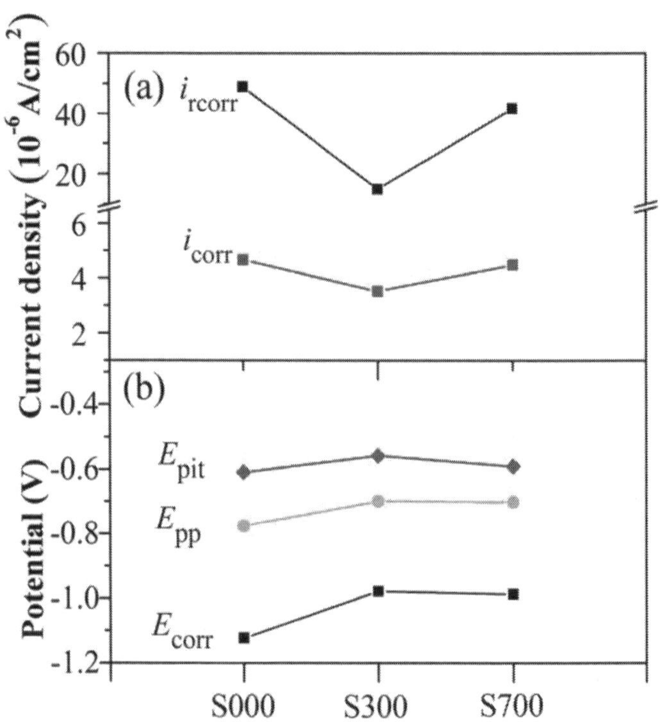

Figure 5. Parameters of the polarization obtained from Figure 4: (a) corrosion current density (icorr) and recorrosion current density ircorr; (b) corrosion potential (Ecorr), pitting potential (Epit) and protection potential (Epp).

Figure 7 shows the weight loss of S000, S300, and S700 samples as a function of sliding time. It is observed that their weight losses increase monotonically with the sliding time, and when the sliding time reaches 7 min, the weight loss of S000 increases more quickly than that of S300 and S700. Overall, the wear resistance of three samples are ranked in the order S300 > S700 > S000, which is consistent with their corrosion resistances (Figure 4 and Figure 6). The SEM micrographs of worn tracks of the three samples are shown in Figure 8. The worn surface of the base metal (S000) shows a large amount of plastic flows. The extent of wear in S300 and S700 is lower than that in S000, i.e. their surfaces are smoother and the groove is shallower compared with S000. The higher magnified view of the wear pits shows that the flaking of the base metal is more serious than with the FSP samples, indicating that the worn debris formation is easier at the base metal (Figure 8b,d,f). In addition, there are a lot of fine particles in the size of several microns around the Fe78Si9B13 particles (Figure 8d,f) and there exist white particles that impede the worn pit extend along the sliding direction (Figure 8c), indicating that the added Fe78Si9B13 particles play a role in improving the samples' wear resistance.

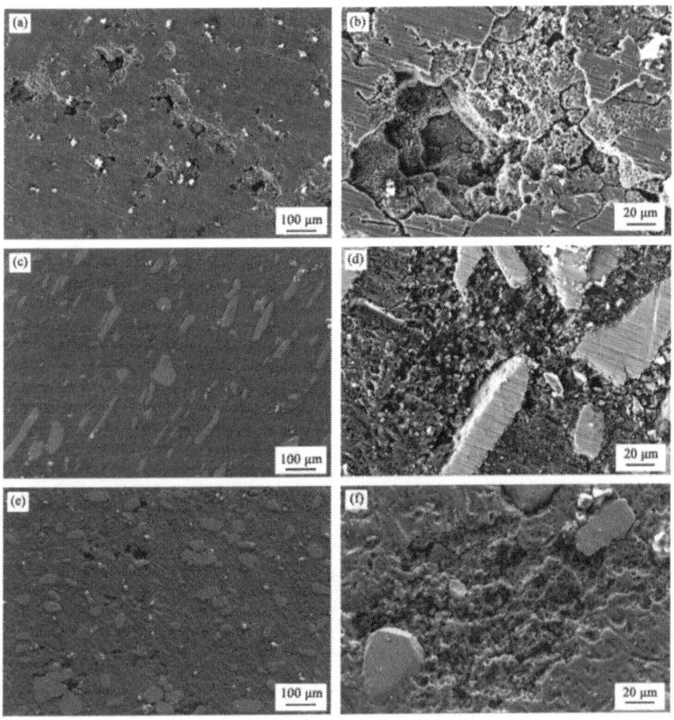

Figure 6. SEM micrographs of electrochemically corroded surfaces: (a and b) S000; (c and d) S300; (e and f) S700.

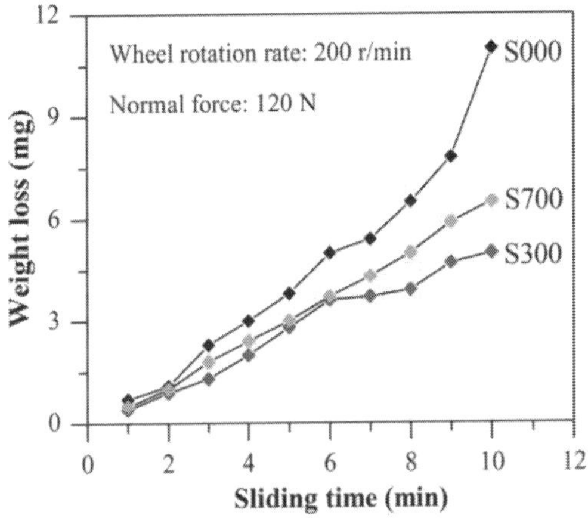

Figure 7. Weight loss of samples as a function of sliding time for base metal (S000) and FSP samples (S300 and S700).

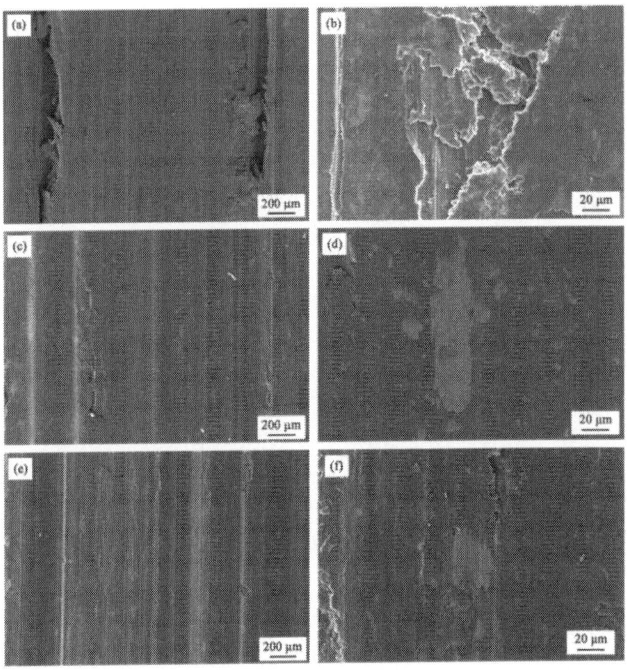

Figure 8. SEM micrographs of worn tracks of samples: (a and b) S000; (c and d) S300; (e and f) S700.

4. DISCUSSION

4.1. Microstructure and Thermal Expansion Behavior of FSP Samples

It is well known that grain refinement in metals and alloys can be realized by friction stir processing (FSP), which was derived based on the principles of friction stir welding (FSW) [9,10] and is similar to the grain refining effect of severe plastic deformation (SPD) [24]. Hence, FSP can be considered one of the variants of the severe plastic deformation (SPD) technologies, which is also discussed in recent reports [25,26]. It has been reported that the SPD treatment accelerates the atomic diffusion inside the processed material, which attributes to the intensive vacancy generation during deformation [27,28,29]. Moreover, the solvus temperature of needle-shape precipitate (β") in 6061 aluminum alloys is about 353 °C, which is lower than the maximal temperature during FSP [30], so the β" can easily dissolve into the matrix during FSP. In addition, the β" mainly contains Mg and Si atoms and the dissolution of Mg and Si into aluminum matrix could increase the lattice constant of α-Al [31], thus it is understood that the lattice parameters of S300 and S700 increase after FSP compared with S000 (Figure 2 inset and Table 1). Furthermore, it is reported that the initial rolling texture is destroyed by the deformation during the FSP [32,33,34]. It is explained that the preferred orientation factor F of the (200) plane in α-Al for S300 and S700 is higher than that of S000 (Figure 2 and Table 1).

In the S300 sample, the Fe78Si9B13 particles annealed at 300 °C are in the amorphous state, while the Fe78Si9B13 particles have crystallized completely during the annealing process at 700 °C (Figure 1b). In the amorphous state, the atoms are metastable and in a random order [19], therefore the Si in Fe78Si9B13 is apt to dissolve into the aluminum matrix and form a solid solution during

FSP. Furthermore, the dissolution of Si could also induce severe lattice distortion and dislocation, which may lead to the dissolution of more Mg and Si into the aluminum matrix. In the S700 sample, the influence of the Fe78Si9B13 particle annealed at 700 °C on the lattice parameter is weaker due to its crystalline structure. As a consequence, the lattice constant a0 of α-Al in S300 is larger than that in S700 (Table 1). Meanwhile, compared with S700, the lower F of S300 indicates a severer recrystallization and then a severer destruction of rolling texture, which may result in a severer lattice distortion and dislocation in S300 and a higher dissolution degree of Mg and Si. This is also confirmed by the fact that the a0 of α-Al in S300 is higher than that in S700 (Table 1).

According to the micro-mechanical model [35], the initial compression stress of a sample induced by CTE mismatch of the constituents is relieved with increasing temperature and causes an increment in the coefficient of thermal expansion (CTE). We determined the initial increment of CTE of the three measured samples (Figure 3). Furthermore, the CTEs for the three samples increase at first, which is also related to the dissolution of the β or β" phase, which mainly contains Mg and Si due to the positive contribution of Mg to CTE and a0 of α-Al solution [31,36,37]. During the cooling process after FSP, the shrinkage degree of base metal was larger than that of the reinforced particles, because the CTE of added particles is less than that of the aluminum matrix alloy, resulting in compressive stress in both the base metal and the added particles [38]. It is understood that the initial CTE of S300 and S700 is lower, and the CTE increment in the beginning stage is higher compared with S000 (Figure 3). Moreover, the maximal CTEs for S300 and S700 are 2.87 × 10−5 K−1 and 3.04 × 10−5 K−1, respectively, and the first maximal of S000 is 2.71 × 10−5 K−1, which is lower than the former values, indicating a higher ability to dissolve Mg for S300 and S700. This is consistent with the a0 variation of measured samples (Table 1). In addition, the temperature at which the CTE begins to decrease for S300 is higher than that for S700, which is due to increased Mg dissolution into the matrix during FSP for S300, which is also reflected in the data of lattice parameters (Table 1). The decrease of CTE for FSP samples after the maximal CTE is related to the Si dissolution into the matrix, supplied by Fe78Si9B13 particles with a negative contribution to CTE of matrix [39]. Moreover, at 550 °C, the CTE of S300 is lower than that of S000 and S700, which is related to shrinkage of the amorphous Fe78Si9B13 particles in S300 crystallized during heat processing.

4.2. Corrosion and Wear Behavior of FSP Samples

After friction stir processing (FSP), the grain size decreased due to dynamic recrystallization [34], which occurs early on in the alloys after SPD [40,41]. The dynamic recrystallization process has undergone a series of paths [42], which involves the dissolution introduction, dynamic recovery, and continuous dynamic recrystallization. It has been reported that the decrease of grain size can reduce the corrosion rate, that is, it improves the corrosion resistance, which is attributed to an ability of fine-grained materials to passivate more readily and to physical breakdown in the second phase of the intermetallic particles that operated as efficient local cathodes and sites for initiation of localized attack [43,44]. Due to the decrease of grain size by dynamic recrystallization during FSP [33,34], the FSP will improve the corrosion resistance of the aluminum alloy, which is in accordance with Rao's research [45]. On the other hand, in the weld head and its vicinity the residual FSP stresses are either parallel or perpendicular to the weld direction; these are compressive stresses due to the CTE mismatch between the base metal and particles [38,46]. The compressive residual stress can improve the corrosion resistance of the surface by reducing the tendency of cracking [34,47,48]. Hence, S300 and S700 have more positive Epit and Epp than S000 and fewer cracks than S000 (Figure 4, Figure 5 and Figure 6). Because the CTE of Fe78Si9B13 in the amorphous state is less than that in the crystallized state [38], the S300 endured larger compressive stress than S700. Moreover, it has been reported that the amorphous samples have a better corrosion resistance than the crystallized ones [49,50]. Hence, it appears that S300 has more positive Ecorr, Epit, and Epp than S700 (Figure 4, Figure 5 and Figure 6).

The presence of added particles converting the wear mode of FSP samples from two bodies to three bodies reduces the loss of wear further by acting as a solid lubricant [5,14,51]. Apparently, the fine powders of the size of several microns around the Fe78Si9B13 particles possibly act as lubricants (Figure 8d,f). In addition, the residual stress after FSP is compressive stress [46], and moderate levels of compressive residual stress have beneficial effects on wear resistance by delaying crack initiation and growth [52]. Furthermore, the smaller grain was conducive to formation of a tribological transformed structure (TTS) layer, and produced shorter delamination cracks in the TTS layer than the larger one [53]—that is, the ultrafine-grained microstructures exhibit markedly enhanced wear resistance relative to the coarse-grained counterpart. Consequently, for the particles located at the end of worn pits, the wear loss of S300 and S700 is lower than S000 (Figure 7). It is known that the amorphous Fe78Si9B13 is metastable and has a drastic stress-induced flow at about 400 °C during the heating process [19,54], which possibly accounts for it acting as a better lubricant (Figure 7). As discussed previously, the difference of CTE for added particles leads to a larger compressive stress in S300 than in S700. Hence, it is shown that S300 has better wear resistance than S700 (Figure 7). In other words, amorphous Fe78Si9B13-reinforced particles can improve the wear and corrosion resistance of AA6061 samples.

5. CONCLUSIONS

The microstructure, thermal expansion behavior, corrosion behavior and, wear resistance of AA6061-T6 with Fe78Si9B13 glass particles by friction stir processing have been investigated (the base metal is denoted as S000, and the FSP samples with the reinforced particles annealed at 300 °C and 700 °C are denoted as S300 and S700, respectively), and it was found that:

(1) By friction stir processing, the Fe78Si9B13 particles are relatively uniform in terms of their dispersion into the Al alloy matrix. The a0 of α-Al for three samples is in the order S300 > S700 > S000, but the F200 for samples is the opposite. During the heating process, the CTEs of the three samples increase firstly and have a maximal value of 2.71 × 10−5 K−1, 2.87 × 10−5 K−1, and 3.04 × 10−5 K−1 for S000, S300, and S700, respectively. The CTE variation can be explained by the dissolving of Mg and Si from the β phase and Fe78Si9B13 particles, as well as the CTE mismatch compressive stress.

(2) The Ecorr and Epit of the three samples are also in the order S300 > S700 > S000, while icorr is quite the opposite. The superior corrosion resistance for S300 can be attributed to the reduction of grain size by FSP and the unique metastable structure of amorphous Fe78Si9B13-reinforced particles. Moreover, the corrosion surface of S300 and S700 has little corrosion pits, but no apparent cracking compared to S000.

(3) The wear resistance of FSP samples has been improved compared with the base metal; in particular, the weight loss for the samples is in the order S300 < S700 < S000. The flaking of S000 after a wear test is more serious than that in S300 and S700. In addition, the worn surfaces of FSP samples are smoother and the grooves shallower compared with S000. The partial improvement of corrosion and wear resistance is ascribed to the compressive stress

in FSP samples due to the CTE mismatch between the base metal and reinforced particles, and the lubrication effect of Fe78Si9B13 particles also plays a role in improving wear resistance.

ACKNOWLEDGMENTS

This work is supported by the National Natural Science Foundation of China (Grants No. 51171091, 51471099), the Excellent Youth Project of Natural Science Foundation of Shandong (No. JQ201012), and the National Basic Research Program of China (973 Program) (No. 2012CB825702).

AUTHOR CONTRIBUTIONS

Lingyu Guo performed the experiments, analyzed the data, and wrote the initial draft of the manuscript; Weimin Wang designed the project, reviewed and contributed to the final revised manuscript; Yan Liu, Kechang Shen, and Kibuem Kim provided helpful feedback and commented on the manuscript; Min Yang and Chaoqun Song performed the experiments and contributed equipment. All authors read the final paper.

REFERENCES

1. Bakes, H.; Benjamin, D. Metals Handbook; ASM international: Metal Park, OH, USA, 1979.
2. Vasudevan, A.K.; Doherty, R.D. Aluminum Alloys-Contemporary Research and Applications: Contemporary Research and Applications; Elsevier: Amsterdam, The Netherland, 2012.

3. Ravi, N.; Sastikumar, D.; Subramanian, N.; Nath, A.K.; Masilamani, V. Microhardness and Microstructure Studies on Laser Surface Alloyed Aluminum Alloy with Ni-Cr. Mater. Manuf. Process. 2000, 15, 395–404.

4. Miracle, D.B. Metal matrix composites–From science to technological significance. Compos. Sci. Technol. 2005, 65, 2526–2540.

5. Shafiei-Zarghani, A.; Kashani-Bozorg, S.F.; Zarei-Hanzaki, A. Microstructures and mechanical properties of Al/Al2O3 surface nano-composite layer produced by friction stir processing. Mater. Sci. Eng. A 2009, 500, 84–91.

6. Budinski, K.G. Surface Engineering for Wear Resistance; Prentice Hall: Upper Saddle River, NJ, USA, 1988.

7. Mabhali, L.; Pityana, S.L.; Sacks, N. Laser Surface Alloying of Aluminum (AA1200) with Ni and SiC Powders. Mater. Manuf. Process. 2010, 25, 1397–1403.

8. Gupta, M.; Mohamed, F.A.; Lavernia, E.J. Solidification behavior of AI-Li-SiCp MMCs processed using variable co-deposition of multi-phase materials. Mater. Manuf. Process. 1990, 5, 165–196.

9. Mishra, R.S.; Mahoney, M.W.; McFadden, S.X.; Mara, N.A.; Mukherjee, A.K. High strain rate superplasticity in a friction stir processed 7075 Al alloy. Scripta Mater. 1999, 42, 163–168.

10. Mishra, R.S.; Ma, Z.Y.; Charit, I. Friction stir processing: A novel technique for fabrication of surface composite. Mater. Sci. Eng. A 2003, 341, 307–310.

11. Ma, Z.Y. Friction stir processing technology: A review. Metall. Mater. Trans. A 2008, 39, 642–658.

12. Aruri, D.; Adepu, K.; Adepu, K.; Bazavada, K. Wear and mechanical properties of 6061-T6 aluminum alloy surface hybrid

composites [(SiC+Gr) and (SiC+Al2O3)] fabricated by friction stir processing. J. Mater. Process. Tech. 2013, 2, 362–369.

13. Clyne, T.W.; Withers, P.J.; Minges, M.L. An Introduction to Metal Matrix Composites; Cambridge University Press: Cambridge, UK, 1995.
14. Ahlatci, H.; Ko Er, T.; Candan, E.; Imeno Lu, H. Wear behaviour of Al/(Al2O3p+SiCp) hybrid composites. Tribol. Int. 2006, 39, 213–220.
15. Liu, Z.Y.; Xiao, B.L.; Wang, W.G.; Ma, Z.Y. Tensile strength and electrical conductivity of carbon nanotube reinforced aluminum matrix composites fabricated by powder metallurgy combined with friction stir processing. J. Mater. Sci. Technol. 2014, 30, 649–655.
16. Makino, A.; Kubota, T.; Chang, C.; Makabe, M.; Inoue, A. FeSiBP bulk metallic glasses with high magnetization and excellent magnetic softness. J. Magn. Magn. Mater. 2008, 320, 2499–2503.
17. Makino, A.; Kubota, T.; Makabe, M.; Chang, C.T.; Inoue, A. FeSiBP metallic glasses with high glass-forming ability and excellent magnetic properties. Mater. Sci. Eng. B 2008, 148, 166–170.
18. Inoue, A. Stabilization of metallic supercooled liquid and bulk amorphous alloys. Acta Mater. 2000, 48, 279–306.
19. Luborsky, F.E. Amorphous Metallic Alloys; Butterworth and Co.: London, UK, 1983.
20. Davis, J.R. Aluminum and Aluminum Alloys; ASM international: Metal Park, OH, USA, 1993.
21. Zhou, Y. Materials Analysis and Testing Technology; Harbin Institute of Technology Press: Harbin, China, 2007.

22. Lotgering, F.K. Topotactical reactions with ferrimagnetic oxides having hexagonal crystal structures—I. J. Inorg. Nucl. Chem. 1959, 9, 113–123.

23. Wilde, B.E.; Williams, E. The use of current/voltage curves for the study of localized corrosion and passivity breakdown on stainless steels in chloride media. Electrochim. Acta 1971, 16, 1970–1985.

24. Azushima, A.; Kopp, R.; Korhonen, A.; Yang, D.Y.; Micari, F.; Lahoti, G.D.; Groche, P.; Yanagimoto, J.; Tsuji, N.; Rosochowski, A. Severe plastic deformation (SPD) processes for metals. CIRP Ann. Manuf. Technol. 2008, 57, 716–735.

25. Nikulin, I.; Malopheyev, S.; Kipelova, A.; Kaibyshev, R. Effect of SPD and friction stir welding on microstructure and mechanical properties of Al-Cu-Mg-Ag sheets. Mater. Lett. 2012, 66, 311–313.

26. Raj, K.H.; Sharma, R.S.; Singh, P.; Dayal, A. Study of friction stir processing (FSP) and high pressure torsion (HPT) and their effect on mechanical properties. Proc. Eng. 2011, 10, 2904–2910.

27. Mazilkin, A.A.; Straumal, B.B.; Rabkin, E.; Baretzky, B.; Enders, S.; Protasova, S.G.; Kogtenkova, O.A.; Valiev, R.Z. Softening of nanostructured Al-Zn and Al-Mg alloys after severe plastic deformation. Acta Mater. 2006, 54, 3933–3939.

28. Straumal, B.B.; Mazilkin, A.A.; Baretzky, B.; Schütz, G.; Rabkin, E.; Valiev, R.Z. Accelerated diffusion and phase transformations in Co-Cu alloys driven by the severe plastic deformation. Mater. Trans. 2012, 53, 63–71.

29. Straumal, B.B.; Gornakova, A.S.; Mazilkin, A.A.; Fabrichnaya, O.B.; Kriegel, M.J.; Baretzky, B.; Jiang, J.Z.; Dobatkin, S.V. Phase transformations in the severely plastically deformed Zr-Nb alloys. Mater. Lett. 2012, 81, 225–228.

30. Sato, Y.S.; Kokawa, H.; Enomoto, M.; Jogan, S.; Hashimoto, T. Precipitation sequence in friction stir weld of 6063 aluminum during aging. Metall. Mater. Trans. A 1999, 30, 3125–3130.

31. Steuwer, A.; Dumont, M.; Peel, M.; Preuss, M.; Withers, P.J. The variation of the unstrained lattice parameter in an AA7010 friction stir weld. Acta Mater. 2007, 55, 4111–4120. A

32. Woo, W.; Choo, H.; Brown, D.W.; Vogel, S.C.; Liaw, P.K.; Feng, Z. Texture analysis of a friction stir processed 6061-T6 aluminum alloy using neutron diffraction. Acta Mater. 2006, 54, 3871–3882.

33. El-Danaf, E.A.; El-Rayes, M.M.; Soliman, M.S. Friction stir processing: An effective technique to refine grain structure and enhance ductility. Mater. Design 2010, 31, 1231–1236.

34. Woo, W.; Balogh, L.; Ungár, T.; Choo, H.; Feng, Z. Grain structure and dislocation density measurements in a friction-stir welded aluminum alloy using X-ray peak profile analysis. Mater. Sci. Eng. A 2008, 498, 308–313.

35. Huber, T.; Degischer, H.P.; Lefranc, G.; Schmitt, T. Thermal expansion studies on aluminium-matrix composites with different reinforcement architecture of SiC particles. Compos. Sci. Technol. 2006, 66, 2206–2217.

36. Uju, W.A.; Oguocha, I.N.A. A study of thermal expansion of Al-Mg alloy composites containing fly ash. Mater. Design 2012, 33, 503–509.

37. Berry, C.R. Effect of internal strains on linear expansion, X-Ray lattice constant, and density of crystals. J. Appl. Phys. 1953, 24, 658–659.

38. Ma, H.J.; Zhang, J.T.; Li, G.H.; Zhang, W.X.; Wang, W.M. Effect of Zr on the thermal stability and magnetic properties of Fe78Si9B13 glassy alloy. J. Alloy. Compd. 2010, 501, 227–232.

39. Zhang, Q.; Wu, G.; Jiang, L.; Chen, G. Thermal expansion and dimensional stability of Al-Si matrix composite reinforced with high content SiC. Mater. Chem. Phys. 2003, 82, 780–785.

40. Straumal, B.B.; Protasova, S.G.; Mazilkin, A.A.; Rabkin, E.; Goll, D.; Schütz, G.; Baretzky, B.; Valiev, R.Z. Deformation-driven formation of equilibrium phases in the Cu–Ni alloys. J. Mater. Sci. 2012, 47, 360–367.

41. Straumal, B.B.; Dobatkin, S.V.; Rodin, A.O.; Protasova, S.G.; Mazilkin, A.A.; Goll, D.; Baretzky, B. Structure and properties of nanograined Fe-C alloys after severe plastic deformation. Adv. Eng. Mater. 2011, 13, 463–469.

42. Su, J.Q.; Nelson, T.W.; Mishra, R.; Mahoney, M. Microstructural investigation of friction stir welded 7050-T651 aluminium. Acta Mater. 2003, 51, 713–729.

43. Ralston, K.D.; Birbilis, N.; Davies, C.H.J. Revealing the relationship between grain size and corrosion rate of metals. Scripta Mater. 2010, 63, 1201–1204.

44. Ralston, K.D.; Birbilis, N. Effect of grain size on corrosion: A review. Corrosion 2010, 66, 1201–1204.

45. Rao, A.G.; Katkar, V.A.; Gunasekaran, G.; Deshmukh, V.P.; Prabhu, N.; Kashyap, B.P. Effect of multipass friction stir processing on corrosion resistance of hypereutectic Al-30Si alloy. Corros. Sci. 2014, 83, 198–208.

46. Staron, P.; KoçAk, M.; Williams, S.; Wescott, A. Residual stress in friction stir-welded Al sheets. Physica B 2004, 350, E491–E493.

47. Van Boven, G.; Chen, W.; Rogge, R. The role of residual stress in neutral pH stress corrosion cracking of pipeline steels. Part I: Pitting and cracking occurrence. Acta Mater. 2007, 55, 29–42.

48. Prevey, P.; Cammett, J. The influence of surface enhancement by low plasticity burnishing on the corrosion fatigue performance of AA7075-T6. Int. J. Fatigue 2004, 26, 975–982.

49. Tam, M.K.; Shek, C.H. Crystallization and corrosion resistance of Cu50Zr45Al5 bulk amorphous alloy. Mater. Chem. Phys. 2006, 100, 34–37.

50. Naka, M.; Hashimoto, K.; Masumoto, T. Corrosion behavior of amorphous and crystalline Cu50Ti50 and Cu50 Zr50 alloys. J. Non-Cryst. Solids 1978, 30, 29–36.

51. Gurcan, A.B.; Baker, T.N. Wear behaviour of AA6061 aluminium alloy and its composites. Wear 1995, 188, 185–191.

52. Mohammadzadeh, R.; Akbari, A.; Drouet, M. Microstructure and wear properties of AISI M2 tool steel on RF plasma nitriding at different N2–H2 gas compositions. Surf. Coat. Technol. 2014, 258, 566–573.

53. Li, J.; Lu, Y.; Zhang, H.; Xin, L. Effect of grain size and hardness on fretting wear behavior of Inconel 600 alloys. Tribol. Int. 2015, 81, 215–222.

54. Niu, Y.C.; Bian, X.F.; Wang, W.M.; Jin, S.F.; Liu, X.J.; Zhang, J.Y.; Qin, G.L. The use of flow behavior and thermal expansion to monitor structural change of amorphous Fe78Si9B13 ribbon. J. Non-Cryst. Solids 2005, 351, 3854–3860.

CHAPTER 10

Development of Self-Healing Coatings Based on Linseed Oil as Autonomous Repairing Agent for Corrosion Resistance

Karan Thanawala [1,2,4,*], Nisha Mutneja [3], Anand S. Khanna [1,2,†] and R. K. Singh Raman [1,4,5,†]

[1] IITB-Monash Research Academy, Powai, Mumbai 400076, India

[2] Department of Metallurgical Engineering and Materials Science, Indian Institute of Technology Bombay, Powai, Mumbai 400076, India

[3] Amity Institute of Nano Technology, Amity University, Noida, Uttar Pradesh 201303, India

[4] Department of Chemical Engineering, Monash University, Victoria 3800, Australia

[5] Department of Mechanical & Aerospace Engineering, Monash University, Victoria 3800, Australia

ABSTRACT

In recent years corrosion-resistant self-healing coatings have witnessed strong growth and their successful laboratory design and synthesis categorises them in the family of smart/multi-functional materials. Among various approaches for achieving self-healing, microcapsule embedment through the material matrix is the main one for self-healing ability in coatings. The present work focuses on optimizing the process parameters for developing microcapsules by in-situ polymerization of linseed oil as core and urea-formaldehyde as shell material. Characteristics of these microcapsules with respect to change in processing parameters such as stirring rate and reaction time were studied by using optical microscopy (OM), scanning electron microscopy (SEM) and Fourier transform infrared spectroscopy (FT-IR). The effectiveness of these microcapsules in coatings was characterized by studying their adhesion, performance, and mechanical properties.

KEYWORDS

linseed oil; self-healing coatings; microcapsules; epoxy coating

1. INTRODUCTION

The corrosion of metals is one of the most destructive processes that causes huge economic losses, in particular, to automotive, marine, oil and gas, and aerospace industries. Some of the efforts taken for its prevention are the use of alternative materials and design of components, and/or application of a suitable protective coating, depending on the type of environmental conditions, the metal is exposed to and the expected life. Among all of these, the most efficient and the most common approach to control corrosion is the application of organic polymer based coatings [1]. However, being the outermost layer on structures, these coatings are susceptible to damage and scratches originating at micro- and nano-levels during handling and service. Such damage is hard to detect, gradually propagating the corrosion process, and finally rendering the coating non-protective. The need is therefore to design and develop coatings which possess the ability to heal the damage thus maintaining the protective properties [2].

An active protection based on “self-healing” of defects in coatings is necessary for a durable effect. Active coatings with self-healing ability can be achieved from the intelligent release systems that are incorporated into the polymer matrix. Application of these coatings is a relatively recent concept in the corrosion protection technology. The active functionality is achieved via incorporation of “smart” release of micro-containers into the polymer. The micro-container can be tailored to trigger release through different mechanisms such as release under mechanical rupture, pH controlled release, ion-exchange controlled release or desorption controlled release [3]. Micro-containers loaded with a variety of functional materials have been used for different application fields such as biotechnology [4,5], preservation of flavors [6,7,8,9], sustainable drug release [9], electro-rheology fluids [10], dyes [11,12], and fire retardant powders [9,13,14].

There are different techniques available for encapsulation of reactive materials, which can be classified on the basis of a wall formation mechanism as reported by Pascault et al. [15]. Also, considerable literature is available on the synthesis methods and parameter optimization for encapsulation. The most important parameters that influence the size, shape, morphology, and thickness of shell of microcapsules are: stirring speed, type of emulsifier, concentration of emulsifier, pH, temperature and the duration of the encapsulation reaction [7,16,17,18,19,20].

In the present study, self-healing coatings, consisting of encapsulated drying oil for corrosion resistance have been investigated. Linseed oil was chosen as healing agent, the significance being its ability to form a film by oxidative drying [17,18,20,21,22,23,24,25,26,27]. However, the drying process can be accelerated by using driers, based on cobalt, calcium, lead, and zirconium. As a basis for further modifications and a useful model reference, the synthesis procedure, proposed by Wang et al. [19] was used. Critical process parameters such as stirring rate and reaction time were optimized for self-healing performance in corrosive environments. Also, adhesion and impact strength were characterized for examining the commercial feasibility of the coating.

2. RESULTS AND DISCUSSION

For self-healing coating containing microcapsules, it is very important for the healing agent to remain intact within the shell, and get released easily when the microcapsules are ruptured. For both properties the shell formed should be thin and compact, which depends on the extent of the polymerization process. To investigate the influence of reaction time on shell thickness and extent of polymerization, the reaction time was varied between 1 and 4 h as shown in Table 1. In the synthesis of microcapsules, formation and stabilization of emulsion is very critical. This is greatly influenced by the concentration of emulsifier and the stirring speed. Thus the stirring speed was varied between 100 and 400 rpm as also shown in Table 1.

The microcapsule mean diameter and shell wall thickness were determined using an optical microscope (Olympus GX51, Olympus Corporation, Shinjuku-ku, Tokyo, Japan) that was equipped with image analyzing software (Olysia m3, Olympus Corporation, Shinjuku-ku, Tokyo, Japan). Features in Figure 1a,b, Figure 2a,b, Figure 3a,b and Figure 4a,b show that the formation of urea-formaldehyde (UF) polymer particles predominated under the respective conditions, and microcapsule formation was negligible. The reason could be the stirring speeds of 100 and 200 rpm, which are insufficient for stabilization of the emulsion (which is also reflected in Table 1), leading to agglomeration of linseed oil forming a separate oil phase. The purpose of stirring is to form stable emulsion by breaking large oil droplets into smaller oil droplets. The agglomeration of oil droplets is favored at low stirring speeds, allowing formation of an unstable emulsion [23,28]. This affects the polymerization process of the urea formaldehyde shell around the oil droplets leading to formation of polymer particles. On the other hand, at stirring speeds of 300 and 400 rpm, the emulsion formed at the same dosage of surfactant was homogeneous and stable. Therefore, the formation of microcapsules was observed as shown in Figure 1c,d, Figure 2c,d, Figure 3c,d and Figure 4c,d.

Table 1. Particle size and shell thickness data with change in reaction time and stirring speed.

Stirring Speed (rpm)	Reaction Time (h)			
	1	2	3	4
100	Particle size: 8 µm Core size: -NA- Shell thickness: -NA-	Particle size: 10 µm Core size: -NA- Shell thickness: -NA-	Particle size: 11 µm Core size: -NA- Shell thickness: -NA-	Particle size: 13 µm Core size: -NA- Shell thickness: -NA-
200	Particle size: 10 µm Core size: -NA- Shell thickness: -NA-	Particle size: 13 µm Core size: -NA- Shell thickness: -NA-	Particle size: 18 µm Core size: -NA- Shell thickness: -NA-	Particle size: 17 µm Core size: -NA- Shell thickness: -NA-
300	Particle size: 11 µm Core size: 6 µm Shell thickness: 5 µm	Particle size: 14 µm Core size: 8 µm Shell thickness: 6 µm	Particle size: 40 µm Core size: 32 µm Shell thickness: 8 µm	Particle size: 45 µm Core size: 15 µm Shell thickness: 30 µm
400	Particle size: 10 µm Core size: 3 µm Shell thickness: 7 µm	Particle size: 12.5 µm Core size: 2.5 µm Shell thickness: 10 µm	Particle size: 31 µm Core size: 21 µm Shell thickness: 10 µm	Particle size: 35 µm Core size: 7 µm Shell thickness: 28 µm

NA: Not applicable.

As can be seen in the Figure 1c,d, Figure 2c,d, Figure 3c,d and Figure 4c,d and the summary in Table 1,the shell thickness increases with increasing reaction time, thereby leading to an increase in the permeability of the shell wall. One of the possible reasons which can be attributed to increased permeability is a facile polymerization reaction for shell formation at the interface in the direction normal to the surface of the core. This facilitated increase in the shell thickness [29]. However, with increase in thickness, the time required to complete cross-linking and polymerization of UF so as to form a compact-nonporous shell, also increases. From Figure 1, Figure 2, Figure 3 and Figure 4 and Table 1, it is evident that the combination of the reaction time of 3 h and the stirring speed of 300 rpm, allows formation of microcapsules with a maximum core having a thin and compact shell wall thickness. Further reaction leads to an increase in shell thickness and permeability.

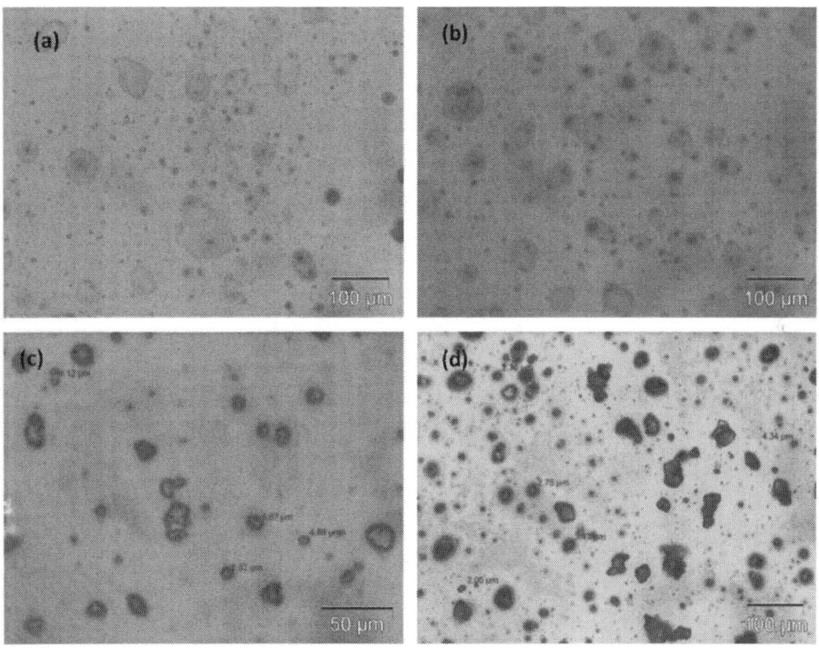

Figure 1. Optical microscopy images of microcapsules developed at a constant reaction time of 1 h but different stirring speeds: (a) 100 rpm; (b) 200 rpm; (c) 300 rpm; (d) 400 rpm.

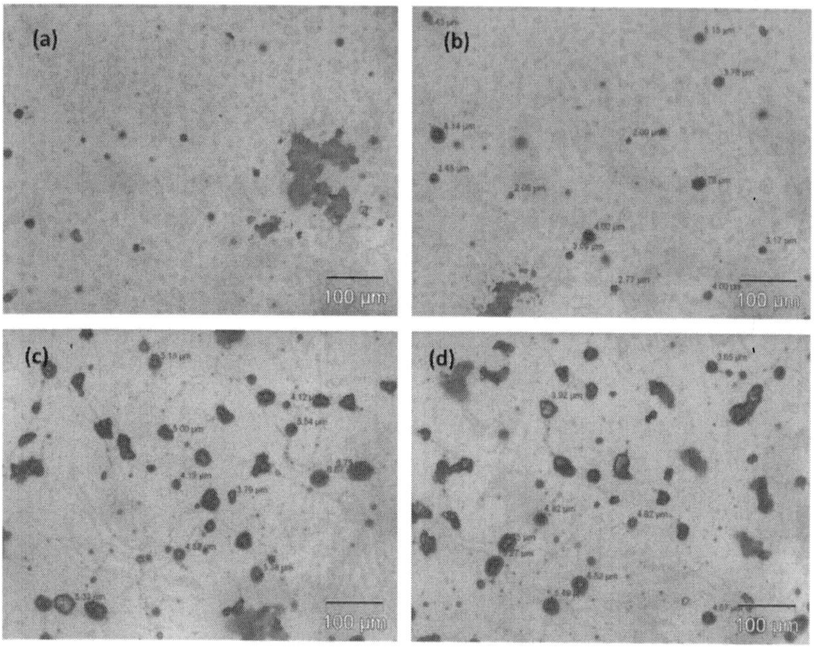

Figure 2. Optical microscopy images of microcapsules developed at a constant reaction time of 2 h but different stirring speeds: (a) 100 rpm; (b) 200 rpm; (c) 300 rpm; (d) 400 rpm.

The surface morphology of the microcapsules obtained at 300 rpm, and reaction time 3 h were investigated by scanning electron microscopy (SEM) (Hitachi S-3400N, High Tech Solutions Corporation, Tokyo, Japan). As can be seen, the microcapsule clearly had a rough outer surface (Figure 5a) and smooth inner surface morphology (Figure 5b). The rough outer surface morphology provides the additional interfacial area necessary for better adhesion with the film matrix. This probably eases the breakage of microcapsules due to stress generated in the scribed area [21].

Microcapsules prepared by an in-situ polymerization method under the optimized conditions were characterized using Fourier transform infrared spectroscopy (Jasco FT/IR 6100, Jasco, Easton, MD, USA) to confirm the presence of linseed oil and UF, the spectrum of UF having characteristic peaks of C–H stretching (2965 cm−1), N–H bending (1535 cm−1), C–H bending (1377 cm−1), C=O stretching (1656 cm−1), C–N stretching (1184 cm−1) and O–H stretching (3280 cm−1). The linseed oil's spectrum having C–H stretching (2933 cm−1), C=C symmetric stretching (1654 cm−1), C=O stretching (1751 cm−1), C–H bending (1456 cm−1) and O–H (2924 cm−1) peaks. All the characteristic peaks of UF and linseed oil were found in the FT-IR spectrum of the microcapsule (Figure 6), which confirms the presence of UF and linseed oil in the microcapsule. Additionally, a shift of the C=C peak from 1654 to 1669 cm−1 of UF and the C=O peak of linseed oil from 1751 to 1741 cm−1 confirms the formation of cross linked polymer.

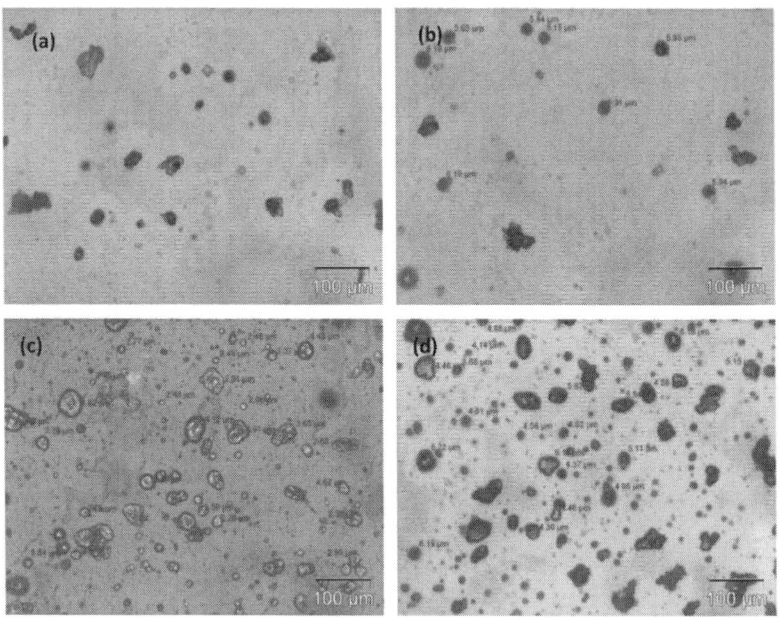

Figure 3. Optical microscopy images of microcapsules developed at a constant reaction time of 3 h but different stirring speeds: (a) 100 rpm; (b) 200 rpm; (c) 300 rpm; (d) 400 rpm.

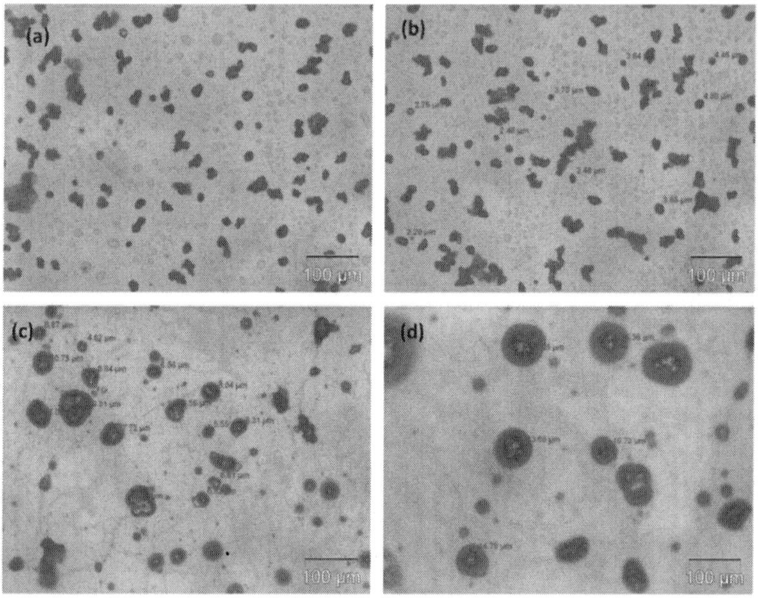

Figure 4. Optical microscopy images of microcapsules developed at a constant reaction time of 4 h but different stirring speeds: (a) 100 rpm; (b) 200 rpm; (c) 300 rpm; (d) 400 rpm.

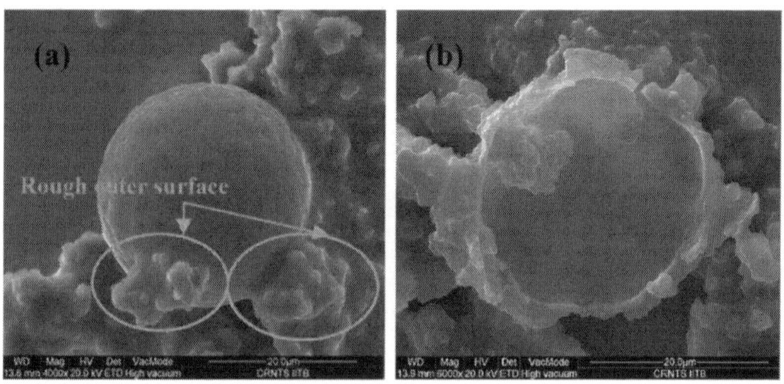

Figure 5. Scanning electron microscopy (SEM) images of microcapsules at 300 rpm and a reaction time of 3 h: (a) rough outer surface and (b) smooth inner surface.

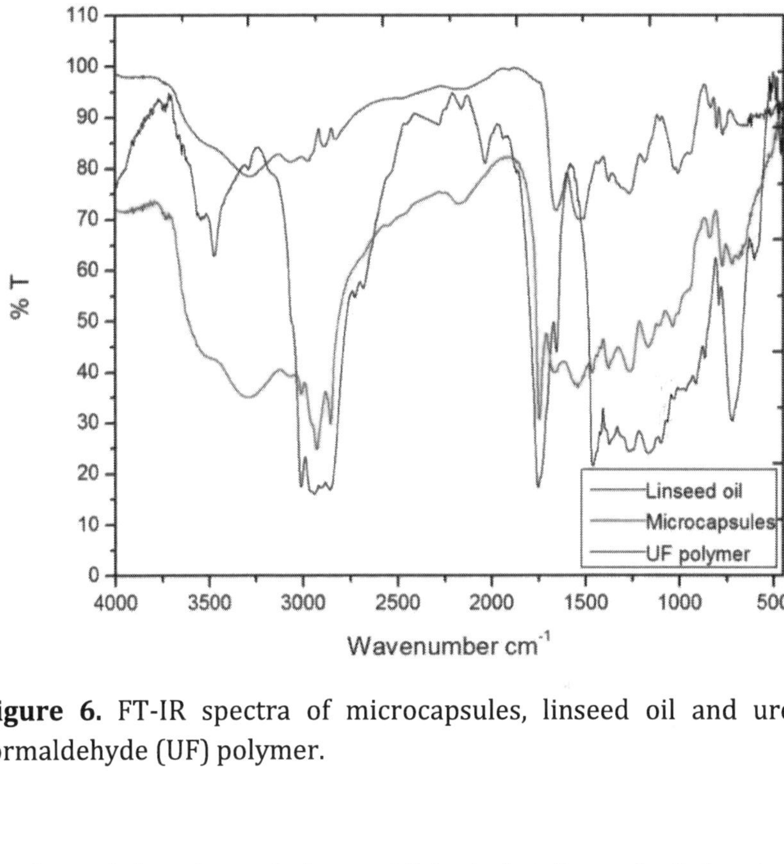

Figure 6. FT-IR spectra of microcapsules, linseed oil and urea-formaldehyde (UF) polymer.

In view of the above, it is established that linseed oil has been successfully encapsulated in the UF shell. The oil content of the prepared microcapsules was determined for an extraction time of 5 h using the Soxhlet process as reported elsewhere [17,18,20,21,22,23,24,25]. The percentage linseed oil content in the microcapsules was determined to be 82% using the Soxhlet process.

2.1. Evaluation of Healing Performance

For preliminary evaluation of the mechanical integrity of the coating, the coating was scribed with the help of a sharp needle, after a curing period of 7 days. The self-healing ability in the coating is provided by linseed oil that is released from the rupturing of the microcapsules, which then fills in the cracks, further forming a film by oxidative polymerization of linseed oil in the presence of atmospheric oxygen and a combination of driers. The healing ability of the coating was examined for 48 h in a laboratory environment. The extent of healing for blank epoxy coating (without embedding microcapsules) and epoxy coating embedded with three different samples with varying microcapsule content (1%, 2%, and 3%) was observed using optical microscopy, as shown in Figure 7, Figure 8 and Figure 9. The coating loaded with 3% microcapsules healed seamlessly within 24 h, while the coating containing 2% microcapsules healed seamlessly in 48 h as shown in Figure 7c,d, Figure 8c,d and Figure 9c,d. In contrast, the coating with 1% microcapsules did not heal completely even after a period of 48 h, due presumably to the lack of healing agent available for seamless integration which is clearly visible from Figure 7b, Figure 8b and Figure 9b. However, it is evident from Figure 7a, Figure 8a and Figure 9a that no healing action can be seen in the blank epoxy coating. The reason can be attributed to the absence of healing agent (linseed oil) encapsulated in the synthesized microcapsules. The SEM Images of artificially scratched self-healing coating embedded with 3% microcapsules, at zero hours and after 48 h are shown in Figure 10. From the Figure 10a,b it can be confirmed that the scratch line had healed successfully. The coating was further characterized for its effectiveness of healing by means of immersion testing in a more corrosive environment.

2.2. Immersion Test

To investigate the effectiveness of healing and the extent of corrosion protection offered by the self-healing coating, the control coating and the coating loaded with 3% microcapsules were immersed in 3.5% NaCl solution. The coatings were manually scribed with the help of a sharp needle to expose the underlying substrate to the corrosive environment. The control coating started forming blisters and rust within 24 h, and accelerated corrosion activity in the coming days. As shown in Figure 11a and Figure 12a, extensive rust formation was observed with increase in the duration of its exposure, particularly in the scribed region of the control coating. In contrast, the coating incorporated with 3% microcapsules showed no visible traces of blistering and rust formation over 200 h immersion (Figure 11b). Even over 500 h of immersion, only slight rust formation was observed in the scribed region (Figure 12b). This remarkable anticorrosive performance of the coating embedded with microcapsules, establishes the excellent ability of oxidative polymerization of linseed oil released from the ruptured microcapsules in healing the cracks, preventing further ingress of oxygen and moisture.

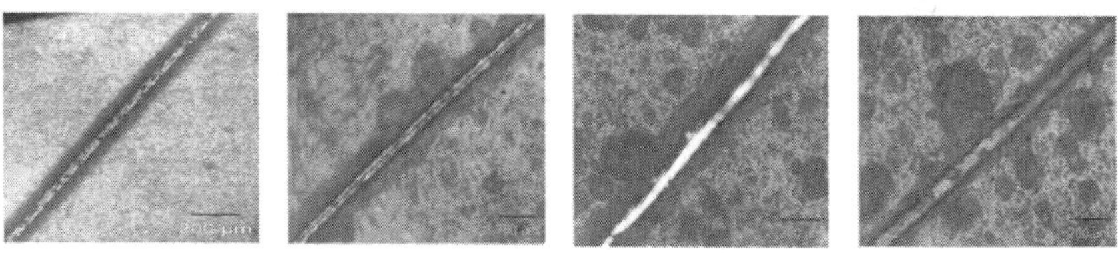

(a) (b) (c) (d)

Figure 7. Self-healing ability of (a) blank epoxy coating and coating embedded with (b) 1 wt%; (c) 2 wt% and (d) 3 wt% of microcapsule (after 0 h).

The adhesion test results (Table 2) show no difference in the mean adhesion values of the control and microcapsule embedded sample. Thus, it is evident that the incorporation of microcapsules did not result in any sacrifice to the adhesion.

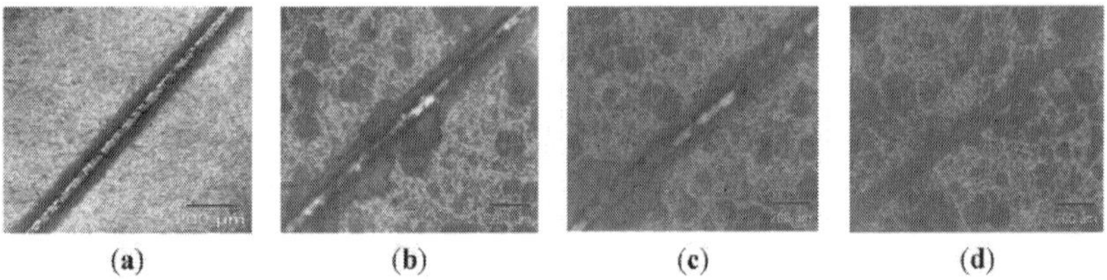

Figure 8. Self-healing ability of (a) blank epoxy coating and coating embedded with (b) 1 wt%; (c) 2 wt% and (d) 3 wt% of microcapsule (after 24 h).

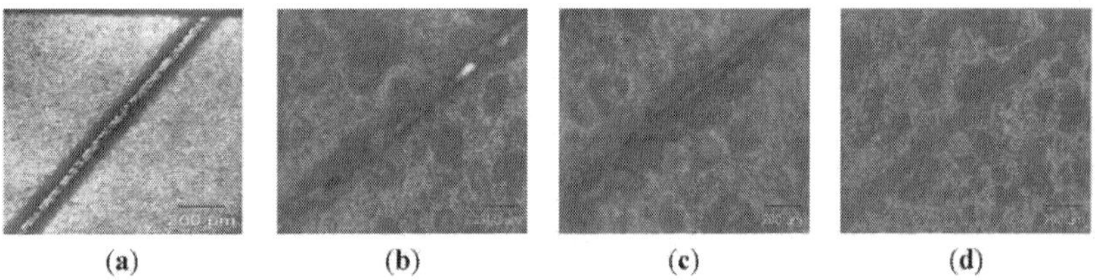

Figure 9. Self-healing ability of (a) blank epoxy coating and coating embedded with (b) 1 wt%; (c) 2 wt% and (d) 3 wt% of microcapsule (after 48 h).

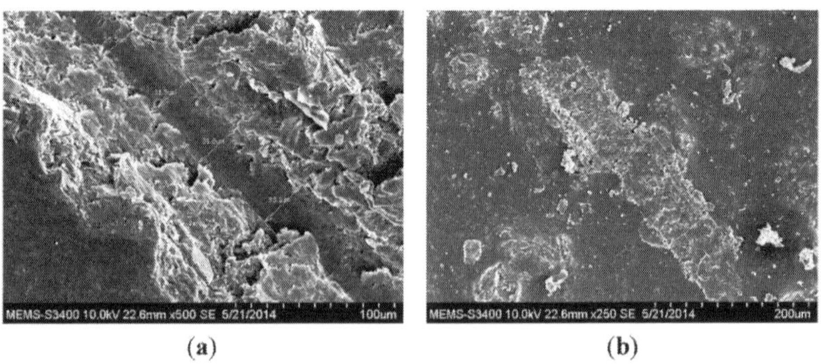

Figure 10. SEM images of the scribed region of the self-healing coating embedded with 3 wt% microcapsules (a) after 0 h of healing time and (b) after 48 h of healing time.

The impact test results in Table 3 indicate no significant change in the resistance offered by the control sample and the coating embedded with microcapsules, and no significant loss of impact resistance.

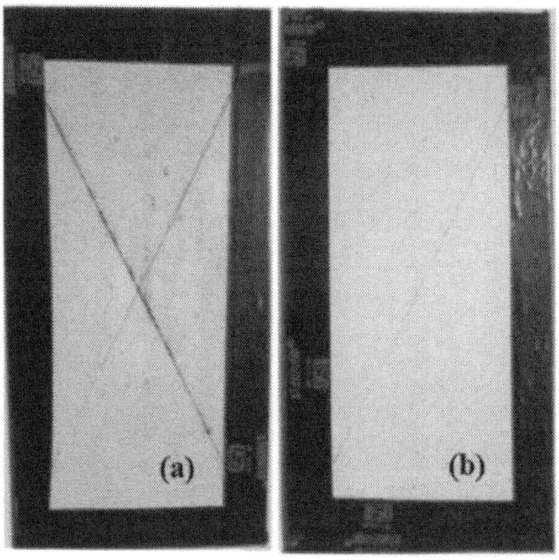

Figure 11. Coating surface after immersion in 3.5% NaCl solution after 200 h; (a) Control coating; (b) Self-healing coating embedded with 3% microcapsules.

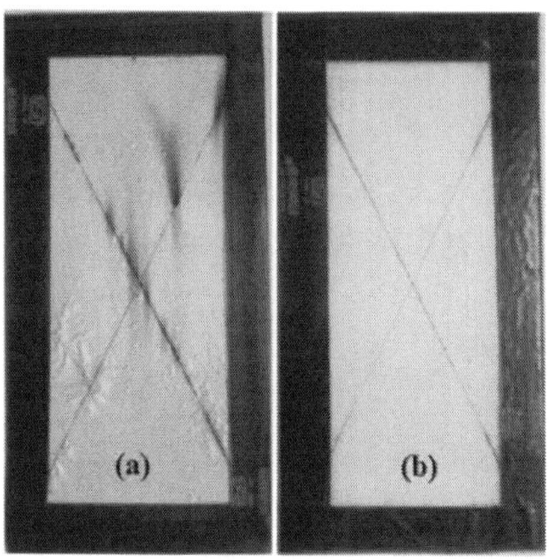

Figure 12. Coating surface after immersion in 3.5% NaCl solution after 500 h; (a) Control coating; (b) Self-healing coating embedded with 3% microcapsules.

Table 2. Adhesion strength of control and microcapsule embedded epoxy coated samples.

Sample	Adhesion strength of samples	Mean Adhesion strength
Control epoxy coating	5B, 5B	5B
Epoxy coating incorporated with microcapsules prepared at 300 rpm for 3 h	5B, 5B	5B

Table 3. Impact strength of control and microcapsule embedded epoxy coated samples.

Sample	Impact Strength of Samples (inch-lbs)	Mean Impact Strength (inch-lbs)
Control epoxy coating	233, 237, 243, 235, 240	237.6
Epoxy coating incorporated with microcapsules prepared at 300 rpm for 3 h	230, 238, 225, 233, 230	231.2

3. EXPERIMENTAL SECTION

The microcapsules were prepared by an in situ polymerization method using an oil-in-water emulsion technique [14,17,18,20,21,22,23,25,27]. The schematic of the process is shown in Figure 13.

3.1. Material

High purity materials required for forming microcapsule wall-forming materials, viz., urea, formaldehyde (37 wt% formaldehyde in water) solution, ammonium chloride, resorcinol, triethanolamine (TEA), hydrochloric acid (HCl) were procured from Merck Co. (Mumbai, India), whereas linseed oil, surface active agent Fynol P, driers cobalt napthenate, zirconium octoate and the epoxy paint were procured from local sources.

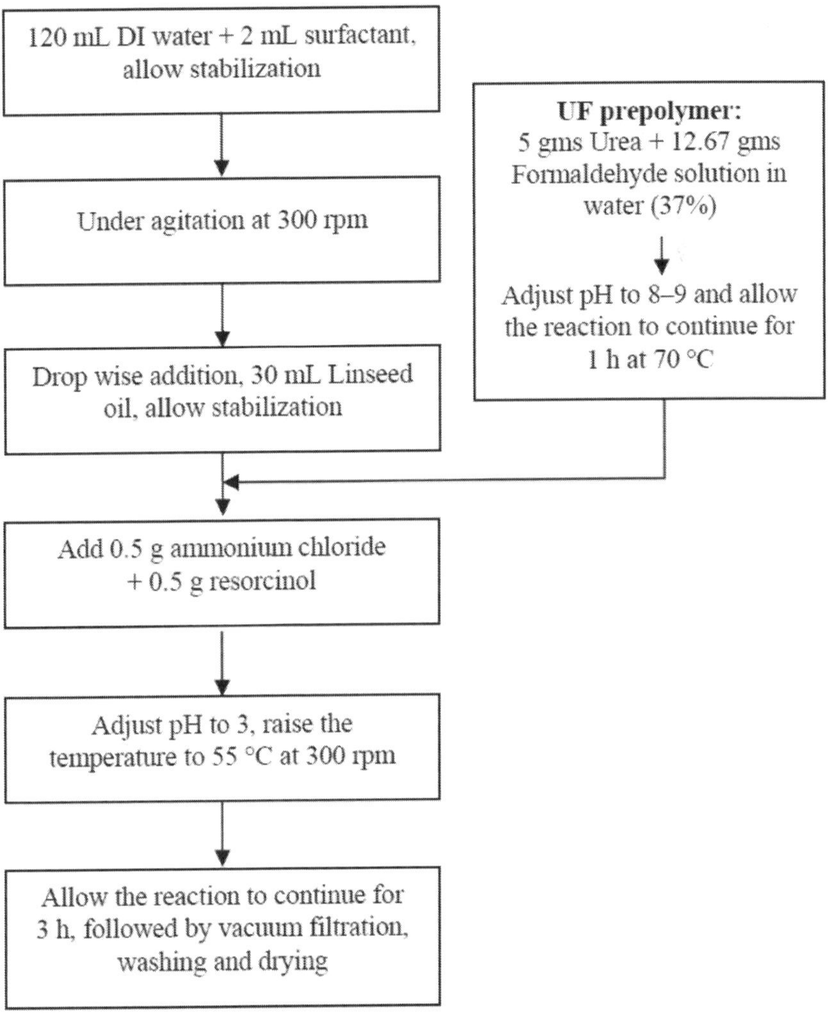

Figure 13. Schematic of the formation of microcapsules by two step in-situ polymerization.

3.2. Preparation of Pre-Polymer

Urea (5 g) and 12.67 g of formaldehyde were mixed in a 100 mL beaker, and stirred at 250 rpm until a clear solution was obtained. The pH of the solution was adjusted to 8–9, with TEA. The solution mixture was heated to 70 °C and held for 1 h, using a water bath. This procedure produced urea formaldehyde (UF) pre-polymer.

3.3. Synthesis of Microcapsules

An optimized mixture of non-ionic and anionic surfactant, Fynol P (2 mL) was added to 120 mL of deionized (DI) water in a 250 mL three necked flask. The mixture was stirred for 15–20 min, and then 25 mL of linseed oil was slowly added to form an emulsion. The emulsion was allowed to stabilize at 100–400 rpm for half an hour. After stabilization, UF pre-polymer was added to the emulsion at room temperature, followed by additions of 0.5 g resorcinol and 0.5 g ammonium chloride. After stirring for 20–30 min, the pH of the emulsion was adjusted to 2–3 by adding 5 wt% HCl. The solution was slowly heated to 50–55 °C, and the reaction was allowed to continue for 1–4 h, before allowing it to cool to ambient temperature. This procedure produced microcapsules that were recovered using vacuum filtration, and then washed thoroughly with DI water and xylene, to remove suspended oil particles and contamination. These microcapsules were vacuum dried at 50 °C. The microcapsules were in the form of a free flowing powder.

3.4. Encapsulation

The mechanism of in-situ polymerization reaction of urea formaldehyde (UF) microcapsules follows two steps. The first step, i.e., addition reaction, involves the reaction between urea and formaldehyde, forming methylol urea as shown in Figure 14a. However, urea being tetra functional, the presence of excess of formaldehyde might lead to formation of tetra methylol urea. To restrict the formation of dimethylol urea, the pH of the mixture was adjusted to 8–9 and a temperature of 70 °C for 1 h. The reaction was carried out for 1 h. The second step involves condensation reaction between the molecules of dimethylol urea forming a low molecular weight pre-polymer. Further reaction leads to cross linked polymer as shown in Figure 14b. The reaction mixture was adjusted to pH 2–3 at 50–60 °C. The reaction speed can be controlled by varying temperature and pH, depending upon the extent of the desired cross-linking [30,31]. The addition and condensation polymerization reaction of urea formaldehyde is shown in Figure 13.

3.5. Determination of Linseed Oil Content in the Microcapsule

The linseed oil content of the microcapsule core was determined using the Soxhlet process [17,18,20,21,22,23,24,25]. The process follows a solvent extraction method, that is used to separate or isolate a species from a mixture of compounds or impurities, and that is based on solubility characteristics. Xylene was used as extracting solvent, owing to its good miscibility with linseed oil. A known weight of microcapsules (Wi) was crushed using a pestle and mortar and transferred to a thimble. Pestle and mortar were rinsed with xylene and added to the thimble. After 5 h of extraction, at approximately 120 °C in an oil bath, the thimble was carefully taken out of the Soxhlet apparatus and after allowing the solvent to drain off completely, it was dried in an oven at 60 °C for 24 h. The final weight of the remained material (Wf) was noted and the core content of synthesized capsules was determined using the following equation:

Percentage linseed oil content of microcapsule (%)=W i −W f W i × 1 00

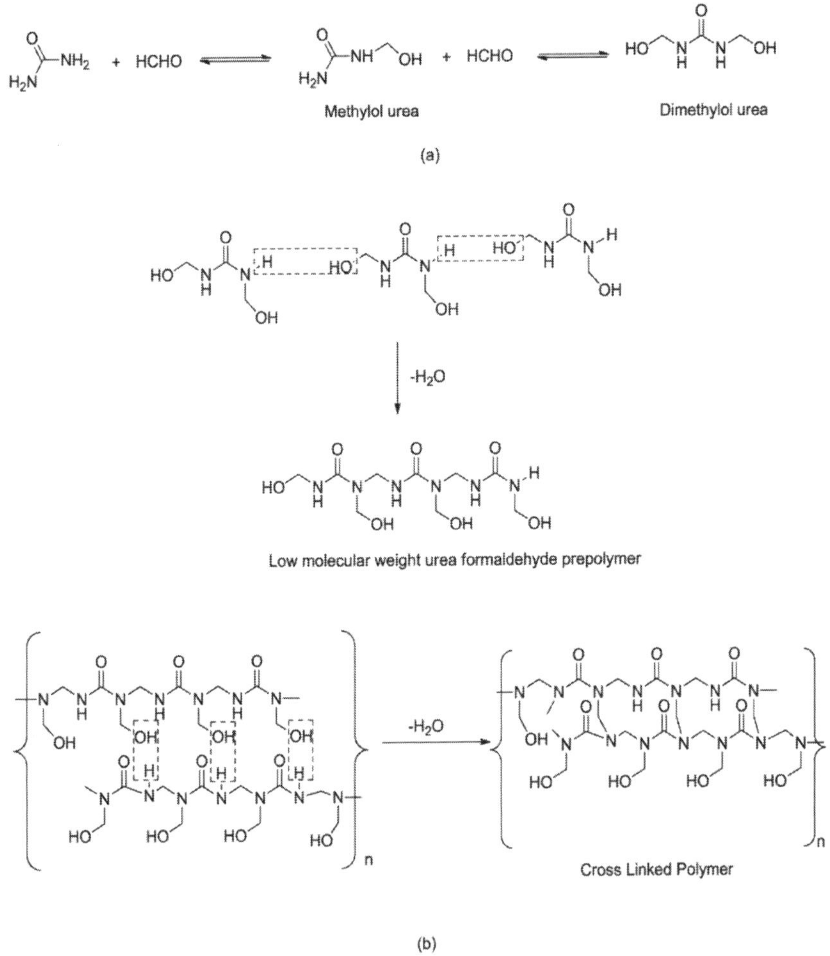

Figure 14. Chemical reaction of urea with formaldehyde form cross linked polymer.

3.6. Preparation of Self-Healing Coating

To investigate the self-healing performance, epoxy based primer system with and without microcapsules was applied on commercially available CR1 grade mild steel substrate (3 inch × 6 inch × 0.0197 inch) as specified in IS 513 [32]. The substrates were first rinsed with acetone to remove contaminants such as oil and dirt, followed by cleaning with 200 mesh size grit paper. The microcapsules were incorporated into the epoxy based primer using two steps. In the first step, microcapsules were allowed to disperse homogeneously in 10 mL of solvent mixture comprising 30% xylene and 70% toluene by means of a water bath ultra sonicator for 30 min. After completion of the dispersion process, the microcapsules were mixed in the primer system using a three-blade mechanical stirrer for 20 min. The prepared self-healing coating was applied using a four-way applicator which gave a uniform coating thickness of around 50–60 microns (measured using Dry Film Thickness Gauge, Elcometer 456, Elcometer Instruments Ltd., Manchester, UK). For evaluation of the mechanical integrity and corrosion resistance of the coating, it was scribed with the help of a sharp needle, after curing for 7 days. The tip radius of the needle was 0.015 mm, made of stainless steel. The coating was scribed at a load of 500 g.

3.7. Adhesion Test

The probable negative effect of microcapsule embedment on the adhesion strength of epoxy coating was investigated by tape test according to ASTM D3359-09 [33]. The test was conducted using two control samples (without microcapsules) and microcapsule embedded coated samples respectively. The assessment involves applying and removing pressure sensitive tape over six cuts made in the film by means of a cross hatch gauge, spaced 2 mm apart.

3.8. Impact Test

Impact tests were conducted to determine the ability of a coating film to resist shattering, cracking, or chipping on rapid deformation, and thus to investigate the negative effect, if any, due to the microcapsule embedment on the mechanical property of the coating. The coated mild steel samples were tested according to ASTM D2794-93 [34], using 0.5 inch indenter diameter and 2 lbs indenter weight.

4. CONCLUSIONS

In the present study, various process parameters were successfully optimized to formulate a self-healing coating using linseed oil as healing material. The most important part of this work was the synthesis of microcapsules, which in turn depends upon various parameters such as stirring rate and reaction time. It was found that a stirring rate of 300 rpm and reaction time of 3 h resulted in microcapsules with optimum core and shell characteristics. The investigations confirmed that an epoxy coating incorporated with 3 wt% microcapsules, gave the best healing characteristics. No significant losses in adhesion and impact strength were identified as a result of the incorporation of microcapsules into the coating. Also, this coating showed remarkably superior resistance to corrosion-assisted blistering during 500 h immersion in 3.5% NaCl, as compared to a control sample having no self-healing constituent.

AUTHOR CONTRIBUTIONS

Karan Thanawala and Nisha Mutneja carried out this work which included paint formulation, synthesis of micro-capsules and characterization of coatings under the guidance of Anand S. Khanna of IIT Bombay and R. K. Singh Raman of Monash University.

REFERENCES

1. Bierwagen, G.P. Corrosion and its control by coatings. In ACS Symposium Series; American Chemical Society: Washington, DC, USA, 1998; pp. 1–8.

2. Huang, M.; Zhang, H.; Yang, J. Synthesis of organic silane microcapsules for self-healing corrosion resistant polymer coatings. Corros. Sci. 2012, 65, 561–566.

3. Zheludkevich, M. Self-healing anticorrosion coatings. In Self-Healing Materials: Fundamentals, Design Strategies, and Applications; Ghosh, S.K., Ed.; Wiley-VCH Verlag GmbH & Co., KGaA: Weinheim, Germany, 2009; pp. 101–139.

4. Orive, G.; Maria Hernández, R.; Rodríguez Gascón, A.; Calafiore, R.; Swi Chang, T.M.; de Vos, P.; Hortelano, G.; Hunkeler, D.; Lacík, I.; Luis Pedraz, J.; et al. History, challenges and perspectives of cell microencapsulation. Trends Biotechnol. 2004, 22, 87–92.

5. Sukhorukov, G.; Fery, A.; Möhwald, H. Intelligent micro- and nanocapsules. Prog. Polym. Sci. 2005, 30, 885–897.

6. Cho, S.H.; White, S.R.; Braun, P.V. Self-healing polymer coatings. Adv. Mater. 2009, 21, 645–649.

7. Liu, X.-D.; Atarashi, T.; Furuta, T.; Yoshii, H.; Aishima, S.; Ohkawara, M.; Linko, P. Microencapsulation of emulsified hydrophobic flavors by spray drying. Dry. Technol. 2001, 19, 1361–1374.

8. Park, S.-J.; Shin, Y.-S.; Lee, J.-R. Preparation and characterization of microcapsules containing lemon oil. J. Colloid Interface Sci. 2001, 241, 502–508.

9. ukaszczyk, J.; Urbaś, P. Influence of the parameters of encapsulation process and of the structure of diisocyanates on the release of codeine from resinate encapsulated in polyurea by interfacial water promoted polyreaction. React. Funct. Polym. 1997, 33, 233–239.

10. Park, B.J.; Lee, J.Y.; Sung, J.H.; Choi, H.J. Microcapsules containing electrophoretic suspension of TiO2 modified with poly(methyl methacrylate). Curr. Appl. Phys. 2006, 6, 632–635.

11. Nelson, G. Application of microencapsulation in textiles. Int. J. Pharm. 2002, 242, 55–62.

12. Sawada, K.; Urakawa, H. Preparation of photosensitive color-producing microcapsules utilizing in situ polymerization method. Dyes Pigments 2005, 65, 45–49.

13. Lee, Y.H.; Kim, C.A.; Jang, W.H.; Choi, H.J.; Jhon, M.S. Synthesis and electrorheological characteristics of microencapsulated polyaniline particles with melamine–formaldehyde resins. Polymer 2001, 42, 8277–8283.

14. White, S.R.; Sottos, N.R.; Geubelle, P.H.; Moore, J.S.; Kessler, M.R.; Sriram, S.R.; Brown, E.N.; Viswanathan, S. Autonomic healing of polymer composites. Nature 2001, 409, 794–797.

15. Pascault, J.-P.; Williams, R.J.J. Epoxy polymers new materials and innovations. In Epoxy Polymers; Wiley-VCH Verlag GmbH & Co., KGaA: Weinheim, Germany, 2010; pp. 1–367.

16. Brown, E.N.; Kessler, M.R.; Sottos, N.R.; White, S.R. In situ poly(urea-formaldehyde) microencapsulation of dicyclopentadiene. J. Microencapsul. 2003, 20, 719–730.

17. Hatami Boura, S.; Peikari, M.; Ashrafi, A.; Samadzadeh, M. Self-healing ability and adhesion strength of capsule embedded coatings—Micro and nano sized capsules containing linseed oil. Prog. Org. Coat. 2012, 75, 292–300.

18. Kouhi, M.; Mohebbi, A.; Mirzaei, M.; Peikari, M. Optimization of smart self-healing coatings based on micro/nanocapsules in heavy metals emission inhibition. Prog. Org. Coat. 2013, 76, 1006–1015.

19. Wang, R.; Li, H.; Hu, H.; He, X.; Liu, W. Preparation and characterization of self-healing microcapsules with poly(urea-formaldehyde) grafted epoxy functional group shell. J. Appl. Polym. Sci. 2009, 113, 1501–1506.

20. Nesterova Tatyana, D.J.; Pedersen, K.; Thorslund, L.; Søren, K. Microcapsule-based self-healing anticorrosive coatings: Capsule size, coating formulation, and exposure testing. Prog. Org. Coat. 2012, 75, 309–318.

21. Suryanarayanaa, C.; Dhirendra Kumara, K.C.R. Preparation and characterization of microcapsules containing linseed oil and its use in self-healing coatings. Prog. Org. Coat. 2008, 63, 72–78.

22. Jadhav, R.; Mane, V.; Bagle, A.; Hundiwale, D.; Mahulikar, P.; Waghoo, G. Synthesis of multicore phenol formaldehyde microcapsules and their application in polyurethane paint formulation for self-healing anticorrosive coating. Int. J. Ind. Chem. 2013, 4, 1–9.

23. Jadhav, R.S.; Hundiwale, D.G.; Mahulikar, P.P. Synthesis and characterization of phenol-formaldehyde microcapsules containing linseed oil and its use in epoxy for self-healing and anticorrosive coating. J. Appl. Polym. Sci. 2011, 119, 2911–2916.

24. Mirabedini, S.M.; Dutil, I.; Farnood, R.R. Preparation and characterization of ethyl cellulose-based core-shell

microcapsules containing plant oils. Colloids Surf. A Physicochem. Eng. Asp. 2012, 394, 74–84.

25. Nesterova, T.; Dam-Johansen, K.; Kiil, S. Synthesis of durable microcapsules for self-healing anticorrosive coatings: A comparison of selected methods. Prog. Org. Coat. 2011, 70, 342–352.

26. Pilbath, A.; Szabo, T.; Telegdi, J.; Nyikos, L. Secm study of steel corrosion under scratched microencapsulated epoxy resin. Prog. Org. Coat. 2012, 75, 480–485.

27. Selvakumar, N.; Jeyasubramanian, K.; Sharmila, R. Smart coating for corrosion protection by adopting nano particles. Prog. Org. Coat. 2012, 74, 461–469.

28. Chen, G.; Tao, D. An experimental study of stability of oil–water emulsion. Fuel Proc. Technol. 2005, 86, 499–508.

29. Song, Y.; Sun, P.; Henry, L.L.; Sun, B. Mechanisms of structure and performance controlled thin film composite membrane formation via interfacial polymerization process. J. Membr. Sci. 2005, 251, 67–79.

30. Singha, A.S.; Thakur, V.K. Synthesis and characterization of S. cilliare fiber reinforced green composites. Int. J. Plast. Technol. 2007, 11, 835–851.

31. Singha, A.S.; Thakur, V.K. Effect of fibre loading on urea-formaldehyde matrix based green composites. Iran. Polym. J. 2008, 17, 861–873.

32. Cold Reduced Low Carbon Steel Sheet and Strip; Indian Standard IS 513; Bureau of Indian Standards: New Delhi, India, 2008.

33. Standard Test Method for Measuring Adhesion by Tape Test; ASTM Standard D3359-09; American Society for Testing and

Materials (ASTM) International: West Conshohocken, PA, USA, 2009.

34. Standard Test Method for Resistance of Organic Coatings to the Effects of Rapid Deformation (Impact); ASTM Standard D2794-93; American Society for Testing and Materials (ASTM) International: West Conshohocken, PA, USA, 2010.

CHAPTER 11

Corrosion Resistance of a Sand Particle-Modified Enamel Coating Applied to Smooth Steel Bars

Fujian Tang [1], Genda Chen [1,*], Richard K. Brow [2] and Michael L. Koenigstein [3]

[1] Department of Civil, Architectural and Environmental Engineering, Missouri University of Science and Technology, Rolla, MO 65401, USA

[2] Department of Materials Science and Engineering, Missouri University of Science and Technology, Rolla, MO 65401, USA

[3] Managing Partner, Pro-Perma Engineered Coatings, Rolla, MO 65401, USA

ABSTRACT

The protective performance of a sand particle-modified enamel coating on reinforcing steel bars was evaluated in 3.5 wt% NaCl solution by electrochemical impedance spectroscopy (EIS). Seven percentages of sand particles by weight were investigated: 0%, 5%, 10%, 20%, 30%, 50% and 70%. The phase composition of the enamel coating and sand particles were determined with the X-ray diffraction (XRD) technique. The surface and cross-sectional morphologies of the sand particle-modified enamel coating were characterized using scanning electron microscopy (SEM). XRD tests revealed three phases of sand particles: SiO_2, $CaCO_3$ and $MgCO_3$. SEM images demonstrated that the enamel coating wetted well with the sand particles. However, a weak enamel coating zone was formed around the sand particles due to concentrated air bubbles, leading to micro-cracks as hydrogen gas pressure builds up and exceeds the tensile strength of the weak zone. As a result, the addition of sand particles into the enamel coating reduced both the coating and corrosion resistances.

KEYWORDS

corrosion resistance; enamel coating; sand particles; EIS; SEM; XRD

1. INTRODUCTION

The most commonly used coatings to protect steel bars from corrosion in reinforced concrete (RC) structures are fusion-bonded epoxy (FBE) and hot-dipped galvanized zinc (HDG) [1,2,3]. The main issue with epoxy-coated bars is the reduction in bond strength or force transmissibility between steel bars and their surrounding concrete [4,5]. As a result, coating imperfection and external damage would cause disbondment of the FBE coating and under-film corrosion [6]. Hot-dip galvanized zinc coatings also have two main concerns in engineering applications. First, the zinc coating corrodes vigorously due to the high alkaline environment in fresh concrete if no passive film is formed. Second, the hydrogen produced in the cathodic reaction increases the porosity of adjacent cement pastes and thus reduces the bond strength between the bars and the concrete [7].

Alternatively, porcelain enamel coatings for reinforcement steel bars have been systematically studied for both enhanced corrosion resistance and increased bond strength in concrete [8,9,10,11,12]. Specifically, three types of enamel coatings (pure, mixed and double enamel) applied on reinforcement steel bars were investigated. The pure enamel is commercially available. The mixed enamel is prepared by mixing 50% of the pure enamel by weight with 50% calcium silicate from Portland cement, which was designed to increase the bond strength between steel bars and concrete by both mechanical (roughness) and chemical (reaction) effects with the surrounding concrete. Results showed that the bond strength was increased up to 15% for coated deformed bars, which was accompanied by the decrease of corrosion resistance due to the change in coating microstructure. These results were confirmed by the observed improvement of a weak layer of transition zone near the steel bars [13]. To improve both corrosion resistance and bond strength, the double enamel was introduced with an inner layer of pure enamel and an outer layer of mixed enamel. Both the pure and double enamels had isolated air bubbles and can prevent the penetration of aggressive chemicals to the surface of steel. The addition of calcium silicate to the enamel, however, produced connected channels in the enamel coating, which provides pathways for aggressive chemicals to reach the steel.

Past studies have shown that adding sands into an epoxy coating can increase the bond strength of an epoxy-coated bar in concrete without affecting its corrosion resistance [14,15]. Therefore, an attempt was made to add sand particles into the enamel coating for simultaneous enhancement of both corrosion resistance and bond strength, since sand particles can increase the enamel coating surface roughness, which would increase the steel bar interlocking with the concrete.

This study aims to investigate the corrosion performance of sand particle-modified enamel coating in 3.5 wt% NaCl solution. Smooth steel bars were coated with enamel that is modified with sand particles of various percentages between 0% and 70% by weight. The surface and cross-sectional morphologies of these coatings were characterized with scanning electron microscopy. The corrosion performance of the coating was evaluated by electrochemical impedance spectroscopy.

2. RESULTS AND DISCUSSION

2.1. Characterization of Enamel Coating and Sand Particles

The phase compositions of the enamel coating and sand particles were analyzed by X-ray diffraction, as presented in Figure 1. An amorphous hump, centered near $2\theta = 27°$, dominates the pattern from the enamel coating, as indicated in Figure 1a. This observation is consistent with the glassy nature of borosilicate materials. Small peaks correspond to the presence of a small amount of crystalline quartz in this sample. As indicated in Figure 1b, three phases of sand particles (SiO2, CaCO3 and MgCO3) were observed.

The cross-sectional and surface morphologies of enamel coatings with different percentages of sand particles are similar. Therefore, only the enamel coating with 70 wt% sand particles was characterized with SEM and presented in Figure 2 for the cross-section of an enamel-coated smooth steel bar. As shown in Figure 2a, the coating thickness ranged from 400 to 700 μm due to the addition of the sand particles. Several small air bubbles (approximately 40 μm in diameter) are observed in the enamel coating and concentrated around the sand particles. These bubbles resulted from the chemical reaction during high temperature enameling [16,17]. They formed a weak zone around the sand particles and were locations for potential micro-cracks. As shown in Figure 2b, all of the sand particles were embedded completely underneath the enamel coating. However, a few micro-cracks were sometimes present around part of the sand particles.

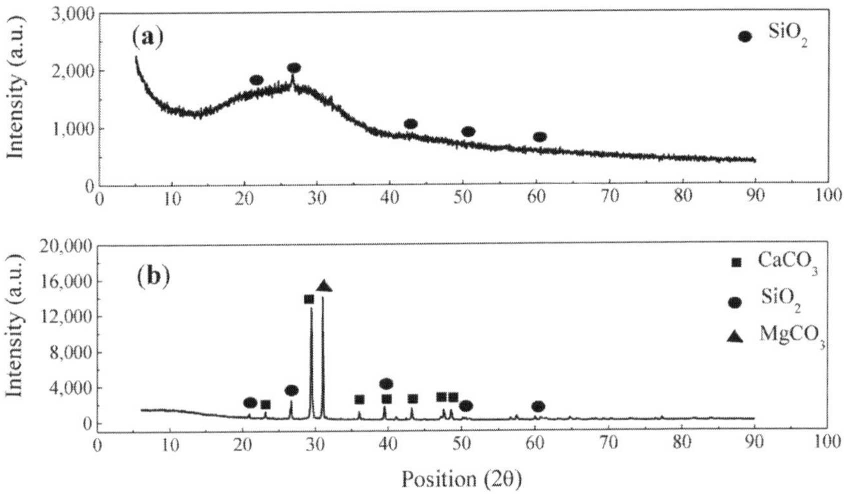

Figure 1. Phase compositions of (a) enamel coating, and (b) sand particles.

Figure 2c,d demonstrates the magnified SEM images of sand particles with and without surrounding micro-cracks, respectively. The presence of micro-cracks around some sand particles is attributed to the increased porosity in the weak zone and the internal hydrogen pressure.

As observed from the XRD analysis, silicate particles (Si02) are the primary constituent of sand particles, although a few carbonate particles (CaC03 and MgC03) are present, as well. Carbonate sand particles are decomposed chemically at high temperature [18,19]. At temperatures between 660 and 979 °C, calcium carbonate breaks down into calcium oxide, with the release of carbon dioxide, as shown in Reaction (1). Similarly, magnesium carbonate is also decomposed between 741 and 838 °C, as shown in Reaction (2). The release of carbon dioxide due to the thermal decomposition of carbonate particles, together with the air bubbles generated in the enameling process, increased the porosity in the weak zone.

$$CaCO_3 \rightarrow CaO + CO_2 \uparrow \quad (1)$$

$$MgCO_3 \rightarrow MgO + CO_2 \uparrow \quad (2)$$

At the enameling temperature, the water in the enamel slurry breaks down into oxygen and hydrogen. Due to its increased solubility in steel, the hydrogen diffuses into the substrate steel at high temperature. However, in the cooling stage, it diffuses out and concentrates at the enamel-steel interface, thus generating internal pressure. Depending upon the amount of hydrogen, the internal pressure at the enamel-steel interface sometimes exceeds the tensile strength of the enamel coating locally, as shown in Figure 2c. As a result, a small piece of enamel coating would flake off the steel substrate. This is the commonly observed fish scale phenomena in enamel coating. In other cases, as shown in Figure 2d, no micro-cracks were observed, likely due to insufficient hydrogen-induced pressure. However, no fish scale was ever observed in the enamel coating without sand particles, since the enamel coating itself is strong enough to resist the internal hydrogen pressure.

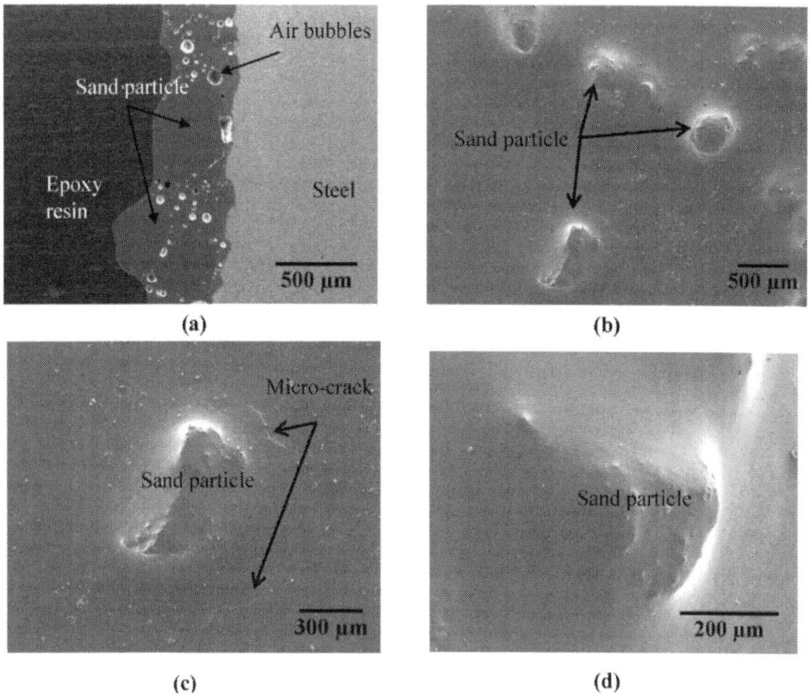

Figure 2. SEM images of enamel coating with 70 wt% sand particles: (a) steel-enamel cross-section; (b) enamel coating surface; (c) sand particle with micro-crack; and (d) sand particle without micro-crack.

2.2. Electrochemical Impedance Spectroscopy

Figure 3 shows the EIS diagrams of uncoated and coated smooth steel bars with sand particle-modified coating in the format of a Nyquist plot. For each percentage of sand particles, three samples were tested with consistent behavior. In these diagrams, the scattered symbols represent the experimental data, and the continuous lines represent the fitted results with equivalent electrical circuit (EEC) models, as will be discussed later. It can be observed from Figure 3 that one large depressed semi-circle appeared in the low frequency range for all samples, including the uncoated steel bars (Figure 3a). This semi-circle is related to the properties of steel-electrolyte interface (double-layer capacitance and charge transfer resistance). With an increase of sand particle percentage, the radius of the large semi-circles decreased significantly. For enamel-coated steel bars with different percentages of sand particles, a small depressed semi-circle appeared in the high frequency range, which is associated with the enamel coating properties (coating capacitance and resistance).

As shown in Figure 4, two EEC models were used to fit the EIS test results for uncoated and enamel-coated steel bars, respectively. These two models were commonly used by other researchers to evaluate the corrosion resistance of steel samples with and without coatings [20,21,22,23,24,25]. Specifically, Rs represents the solution resistance, Rc and CPEc denote the coating resistance and capacitance and Rct and CPEdl represent the charge transfer resistance and double-layer capacitance at the steel-electrolyte interface.

Replacement of capacitance with a constant phase element (CPE) in the EEC models is attributed to the non-homogeneity in the corrosion system, which is generally attributed to the distributed surface reactivity, roughness, electrode porosity and current and potential distribution associated with the electrode geometry [26]. The impedance of CPE can be represented by the following equation:

$$Z_{CPE} = 1/[Y(j\omega)^n] \quad (3)$$

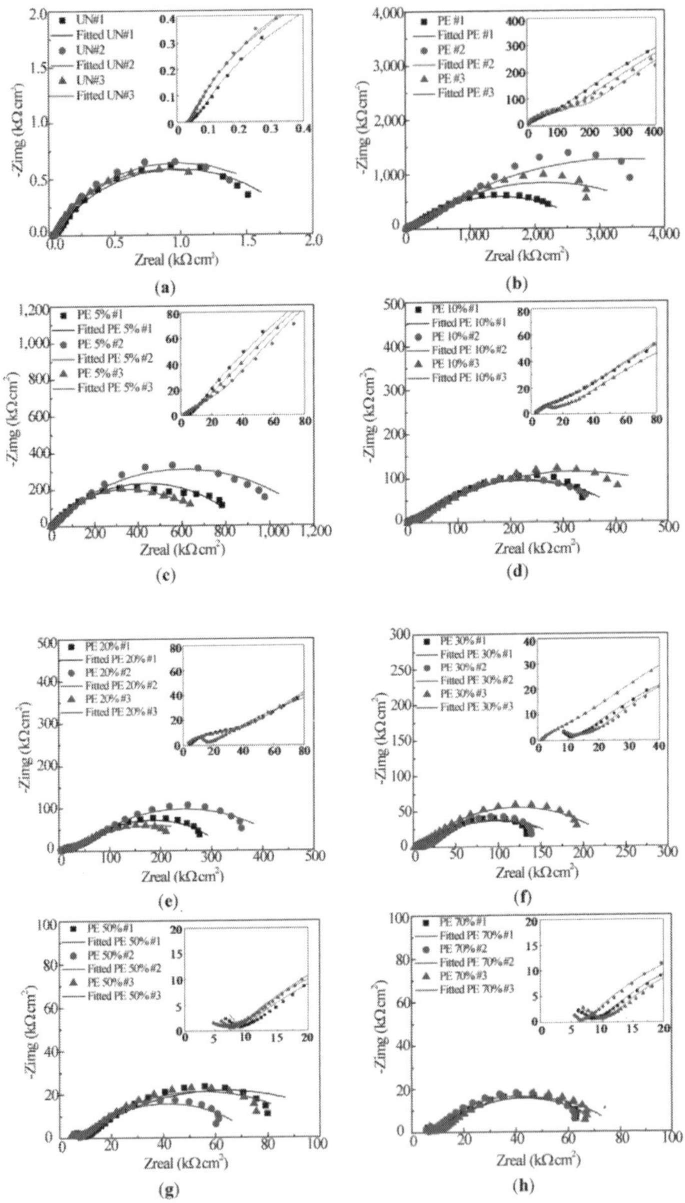

Figure 3. Typical EIS diagrams of smooth steel bars: (a) uncoated (UN); and (b–h) coated with pure enamel (PE) modified with 0%, 5%, 10%, 20%, 30%, 50% and 70% sand particles by weight.

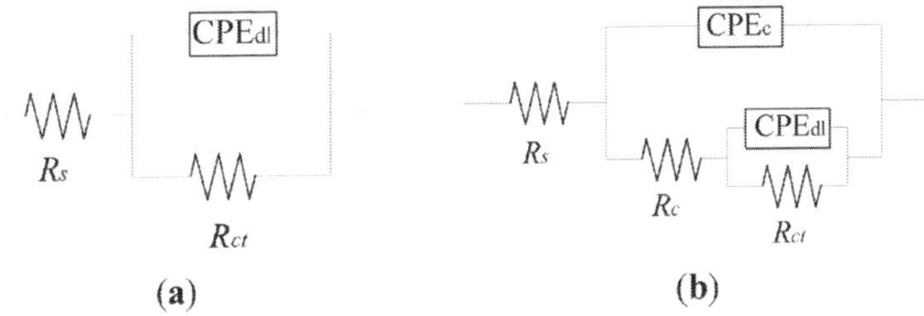

Figure 4. Equivalent electrical circuit model for the (a) uncoated and (b) enamel-coated steel bar. CPE, constant phase element.

where Y and n are two parameters related to the CPE. When n = 1, CPE resembles a capacitor with capacitance Y. When n = 0, CPE represents a resistor with resistance Y−1. The effective capacitance can be calculated from the CPE parameters in a normal distribution non-homogeneity condition [27]:

$$C = Y^{1/n} R^{(1-n)/n} \quad (4)$$

where parameters Rc, Yc and nc were used to calculate the effective capacitance of the coatings Cc; Rct, Ydl and ndl were used to calculate the effective capacitance of double-layer Cdl, respectively.

ZSimpWin was used to fit the EIS data with the two EEC models. The chi-squared values in the fitting are in the range from 10−4 to 10−3, indicating a good fitting. The fitted results were shown in Figure 3 by the continuous line.

Figure 5 shows the dielectric properties of sand particle-modified enamel coatings in terms of coating resistance Rc and coating capacitance Cc. Each represents the average of three samples with an error bar representing one standard deviation. In general, the coating resistance is attributed to the electrical resistance to ionic transfer through the coating pores, which reflects the anti-penetrating ability of the coating to the electrolyte [28]. It is closely related to the dielectric properties, microstructure, thickness and defects of the enamel coatings modified by sand particles. As shown in Figure 5a, the coating resistance decreased from 83.2 ± 42.5 kΩ·cm2 for enamel coating without sand particles to 8.98 ± 1.53 kΩ·cm2 for enamel coating with 70 wt% sand particles. The reduction in coating resistance is due to the presence of micro-cracks around some sand particles. The coating resistance decreased rapidly when the sand particles were lower than 50%. However, it remained stable when the sand particles were equal to or over 50%. This is because when the sand particle content reached 50%, the sand particle area over the entire surface area of enamel coating reached over 90%; the addition of more sand particles did not further change the surface morphology significantly. Therefore, the area with micro-cracks would be similar, resulting in the stabilized corrosion performance.

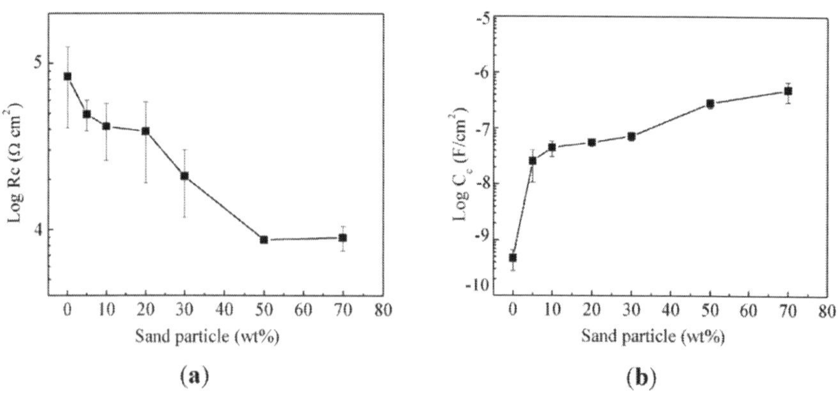

Figure 5. The effect of sand particle content on coating properties: (a) coating resistance; and (b) coating capacitance.

On the contrary, as shown in Figure 5b, the coating capacitance increased with an increase of sand particle content from 0.47 ± 0.19 nF/cm2 for steel bar samples without sand particles to 0.49 ± 0.19 μF/cm2 for steel bars with 70 wt% sand particles. The coating capacitance is directly proportional to the area of the capacitor [29]. The relatively lower capacitance for steel bars without sand particles indicates a comparatively lower exposure of steel substrate to the electrolyte at the pinholes of the enamel coating, as observed in Figure 8a. As more sand particles were added, more micro-cracks were introduced, resulting in the penetration of more NaCl solution into the coating (water uptake), and the coating capacitance increased consequently.

Figure 6 shows the evolution of charge transfer resistance Rct and effective double-layer capacitance Cdl with an increase of sand particle content. The charge transfer resistance measures the ease of electron transfer across the metal surface, which is inversely proportional to corrosion rate [28]. As indicated in Figure 6a, the charge transfer resistance decreased from 4.42 ± 1.78 MΩ·cm2 for enamel-coated bars without sand particles to 67.7 ± 2.00 kΩ·cm2 for enamel coated bars with 70 wt% sand particles. Correspondingly, the double-layer capacitance increased from 0.28 ± 0.12 to 28.6 ± 6.8 μF/cm2, as indicated in Figure 6b. The decrease of charge transfer resistance and the increase of double-layer capacitance with an increase of sand particle content are attributed to the enlargement of the active corrosion area due to presence of more micro-cracks.

Compared with enamel coating modified by calcium silicate (Portland cement) in the previous study [10], the sand particle-modified enamel coating in this study has better performance in terms of both coating and corrosion properties. For the calcium silicate-modified enamel coating, the coating resistance, the coating capacitance, the charge transfer resistance and double-layer capacitance are 0.14 ± 0.05 kΩ·cm2, 15.2 ± 6.4 μF/cm2, 18.3 ± 4.6 kΩ·cm2 and 635 ± 184 μF/cm2, respectively. The addition of calcium silicate into the enamel changes its microstructure from a solid coating layer with isolated air bubbles into an amorphous and porous structure with interconnected channels [9]. Although it is 10-times higher than the corrosion resistance of uncoated bars (Rct = 1.93 ± 0.13 kΩ·cm2), the corrosion resistance of steel bars with calcium silicate-modified enamel coating is 3.7-times lower than that of the steel bars coated with sand particle-modified enamel when the sand particle content reaches 70 wt%. Similarly, the coating resistance of sand particle-modified enamel with 70 wt% sand content is 64-times higher than that of the calcium silicate-modified enamel. Therefore, sand particles as an additive to enamel coating are more desirable than calcium silicate.

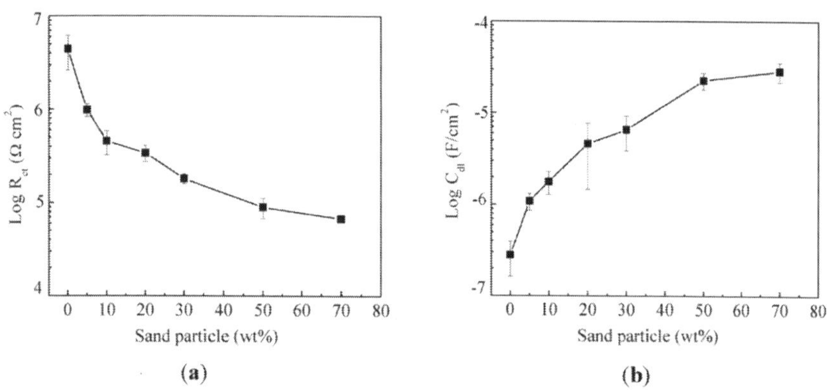

Figure 6. The effect of sand particle contents on: (a) charge transfer resistance; and (b) double-layer capacitance.

Figure 7 shows the effect of sand particles on the non-homogeneities of both coating and a steel-electrolyte interface in terms of coating index nc and double-layer index ndl. The coating index nc decreased from 0.75 for steel bars without sand particles to 0.50 for steel bars with sand particles. The non-homogeneity of enamel coating without sand particles was attributed to the pinhole and coating defect generated in the enameling process. The decrease of nc after the addition of sand particles was attributed to the presence of micro-cracks and the different types of sand particles, both of which reduced the homogeneity of the enamel coating. However, no general trend was observed for the evolution of steel-interfacial index ndl for enamel-coated steel bars with or without sand particles. The value of ndl fluctuated around 0.50.

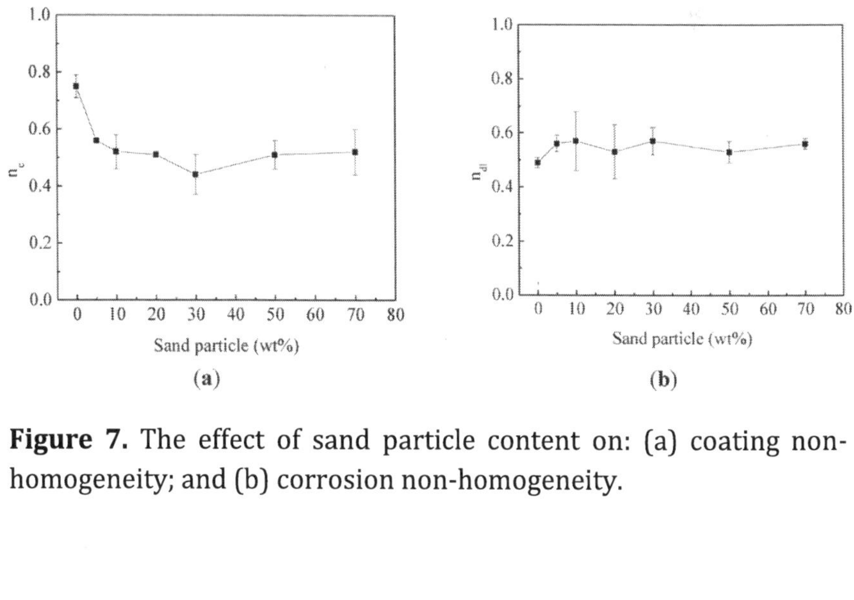

Figure 7. The effect of sand particle content on: (a) coating non-homogeneity; and (b) corrosion non-homogeneity.

2.3. Visual Observation and Analysis

Figure 8 shows the surface conditions of all steel bar samples after having been immersed in 3.5 wt% NaCl solution for seven days. Most of the corrosion products accumulated around some sand particles with micro-cracks. For enamel-coated steel bars without sand particles, corrosion products could be observed at the pinholes of the coating, as shown in Figure 8a. For other enamel-coated steel bars with sand particles from Figure 8b–g, the active corrosion sites increased with an increase of sand particle content. The visually observed conditions of the enamel-coated steel bars with 50% and 70% sand particles were similar. This is because the sand particles covered most of the enamel coating area with similar active corrosion sites. This observation is consistent with the EIS results.

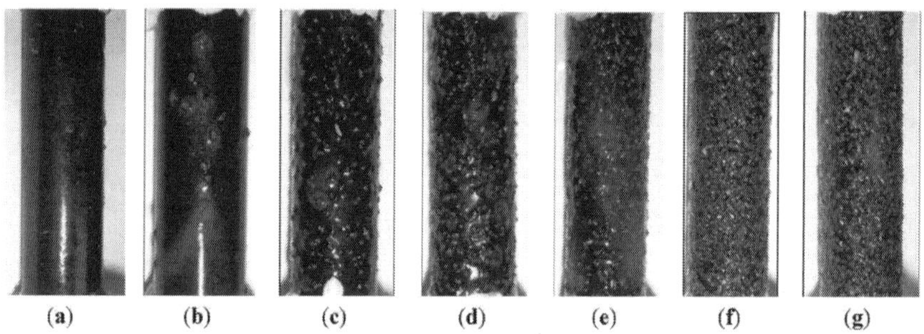

Figure 8. Surface conditions of enamel-coated steel bars after seven days of immersion in 3.5 wt% NaCl solution: (a) 0%; (b) 5%; (c) 10%; (d) 20%; (e) 30%; (f) 50%; and (g) 70% sand particles.

3. EXPERIMENTAL SECTION

3.1. Preparation of Enamel Coating and Samples

Enamel is a glass obtained by fusion at high temperature, and it may be applied to steel using either a wet or dry process. In this study, enamel coating was applied to the surface of steel bars with the wet process. Commercially-available alkali borosilicate glass frits from PEMCO (Product No. PO2025, PEMCO International, Leesburg, AL, USA) were directly used as pure enamels, and their chemical composition is 44.0 SiO2, 19.3 B2O3, 15.8 Na2O, 2.8 K2O, 4.7 CaF2, 4.6 Al2O3, 5.3 ZrO2, 0.9 CoO, 1.5 MnO2, 1.0 NiO (in weight percent) [30]. Enamel slurry was prepared by first adding 454 kg of enamel frits to 189.3 liters of water and mixing them for 20 min and, then, adding clay (31.8 kg) and borax (2.3 kg) as suspension agents and mixing again for 3.5 h. To get various sand particle contents in pure enamel, seven percentages by weight were considered: 0%, 5%, 10%, 20%, 30%, 50% and 70%.

The sand particles used are manufactured sands, whose diameter ranges from 300 to 600 μm with a median particle size of 400 μm (30–50 mesh particle size). To characterize the chemical composition, a 5.0-g sand particle was grounded into powder and analyzed with X-ray diffraction analysis to determine its phase composition.

Grade 60 smooth reinforcing steel bars with a diameter of 13 mm were used, and their chemical composition is 0.43 C, 0.22 Si, 0.95 Mn, 0.15 P, 0.07 S, 0.17 Cr, 0.03 Mo, 0.10 Ni, 0.01 Co, 0.46 Cu, 0.02 V, 0.02 Sn, and Fe balance (in weight percent). Each steel bar sample was sectioned into 76.3 mm-long pieces and then cleansed with a water-based solvent. For the coating process, the steel bar was dipped into the enamel slurry with different sand particle percentages and then heated for 2 min at 150 °C to drive off moisture. The enamel-coated steel bar was then heated in a gas-fired furnace to 810 °C for 10 min and finally cooled to room temperature. The firing treatment at high temperature melts the glass frit and fuses the enamel to the steel.

After coating, one end of the steel bar sample was first polished to expose the substrate steel on which a copper wire was welded for the corrosion test setup. PVC tubes containing epoxy resin were then used to cover the two ends in order to limit the middle portion of the steel bar being exposed to the corrosive environment, as shown in Figure 9a. Therefore, the actual length of the steel potentially exposed to the corrosive environment was approximately 38.1 mm long, and the surface area was approximately 15.2 cm2. A representative steel bar sample with 70 wt% sand particle-modified enamel coating is shown in Figure 9b. Steel bar samples without enamel coting were also prepared for comparison. Three identical samples were prepared for each condition to ensure the repeatability of test results.

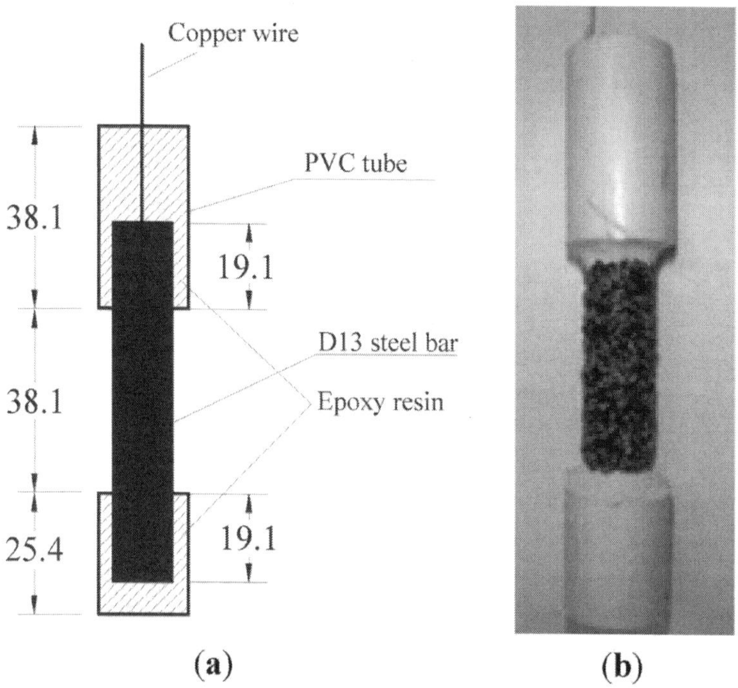

Figure 9. Geometry of steel bar samples: (a) sample dimensions; and (b) sample ready for test (unit: mm).

3.2. Characterization of Enamel Coating with Sand Particles

The microstructures of enamel coatings were investigated with scanning electron microscopy (SEM, Hitachi S4700, Hitachi Ltd., Tokyo, Japan). Two types of enamel-coated steel bar with 70 wt%

sand particles were prepared to understand both the surface and the cross-sectional morphologies with SEM images. For the observation of surface morphologies, a 3.0 cm-long steel bar was used. For the observation of the cross-section, a 1.0 cm-long steel bar was sliced from the bar with 70 wt% sand particles. These samples were cold-mounted in epoxy and ground with silicon carbide papers with grits of 80, 180, 320, 600, 800 and 1200. The samples were rinsed with deionized water, cleansed with acetone and finally dried in air at room temperature. To avoid sample charging, a carbon coating was applied prior to SEM analysis.

3.3. Electrochemical Tests

All samples were immersed to 3.5 wt% NaCl solution. The solution was made by mixing purified sodium chloride with deionized water. The pH of the solution was 5.72 at room temperature. A typical three-electrode set-up was used for the EIS tests. A 25.4 mm × 25.4 mm × 0.254 mm platinum sheet functioned as a counter electrode, a saturated calomel electrode (SCE) as a reference electrode and the steel bar sample as the working electrode. All three electrodes were connected to a Gamry (Gamry Instruments, Warminster, PA, USA). Reference 600 potentiostat/galvanostat with zero resistance ammeter (ZRA), for data acquisition. The tests were conducted at five points per decade around the open-circuit potential Eocp with a sinusoidal potential wave of 10 mV in amplitude and frequency ranging from 100 kHz to 5 MHz. After 7 days of immersion tests, all samples were taken out of the solution for visual observation of the corroded surface.

4. CONCLUSIONS

The corrosion resistance of steel bars with enamel coatings modified by different percentages of sand particles was tested in 3.5 wt% NaCl solution by EIS. The phase composition and microstructure were characterized with X-ray diffraction and scanning electron microscopy, respectively. Based on the test results and data analysis, the following conclusions can be drawn:

(1) Enamel coating wetted well with sand particles and steel substrates. However, a weak zone was formed around the sand particles due to concentrated air bubbles.

(2) The presence of micro-cracks around some sand particles was mainly because the internal pressure of released hydrogen gas exceeded the tensile strength of the enamel coating in the weak zone.

(3) The addition of sand particles reduced both the coating resistance and the corrosion resistance of enamel-coated steel bars. The corrosion resistance of enamel-coated steel bars first decreased rapidly with the increasing percentage of sand particles and then remained stable after 50% sand particles by weight have been added.

ACKNOWLEDGMENTS

The authors gratefully acknowledge the financial support provided by the U.S. National Science Foundation under Award No. CMMI-0900159. The enamel-coated specimens tested in this study were prepared by Pro-Perma Engineered Coatings, Rolla, MO, USA.

AUTHOR CONTRIBUTIONS

Fujian Tang conducted all of the experimental work and drafted the manuscript. Genda Chen provided background information, analyzed and discussed the corrosion test results and finalized the manuscript. Richard K. Brow analyzed and discussed the SEM and XRD results. Michael L. Koenigstein coated the steel bar samples with enamel and sand particles.

REFERENCES

1. Kobayashi, K.; Takewaka, K. Experimental studies on epoxy coated reinforcing steel for corrosion protections. Int. J. Cem. Compos. Lightweight Concr. 1984, 6, 99–116.

2. Darwin, A.B.; Scantlebury, J.D. Retarding of corrosion processes on reinforcing bar in concrete with an FBE coating. Cem. Concr. Compos. 2002, 24, 73–78.

3. Sistonen, E.; Cwirzen, A.; Puttonen, J. Corrosion mechanism of hot-dip galvanized reinforcement bar in cracked concrete. Corros. Sci. 2008, 50, 3416–3428.

4. Treece, R.A.; Jirsa, J.O. Bond strength of epoxy-coated reinforcing bars. ACI Mater. J. 1989, 86, 167–174.

5. Miller, G.G.; Kepler, J.L.; Darwin, D. Effect of epoxy coating thickness on bond strength of reinforcing bars. ACI Struct. J. 2003, 100, 314–320.

6. Manning, D.G. Corrosion performance of epoxy-coated reinforcing steel: North American experience. Constr. Build. Mater. 1996, 10, 349–365.

7. Tan, Z.Q.; Hansson, C.M. Effect of surface condition on the initial corrosion of galvanized reinforcing steel embedded in concrete. Corros. Sci. 2008, 50, 2512–2522.

8. Tang, F.; Chen, G.; Volz, J.S.; Brow, R.K.; Koenigstein, M.L. Cement-modified enamel coating for enhanced corrosion resistance of steel reinforcing bars. Cem. Concr. Compos. 2013, 35, 171–180.

9. Tang, F.; Chen, G.; Brow, R.K.; Volz, J.S.; Koenigstein, M.L. Corrosion resistance and mechanism of steel rebar coated with three types of enamel. Corros. Sci. 2012, 59, 157–168.

10. Tang, F.; Chen, G.; Volz, J.S.; Brow, R.K.; Koenigstein, M.L. Microstructure and corrosion resistance of enamel coatings applied to smooth reinforcing steel. Constr. Build. Mater. 2012, 35, 376–384.

11. Yan, D.; Reis, S.; Tao, X.; Chen, G.; Brow, R.K.; Koenigstein, M.L. Effect of chemically reactive enamel coating on bonding strength that steel/mortar interface. Constr. Build. Mater. 2012, 28, 512–518.

12. Wu, C.; Chen, G.; Volz, J.S.; Brow, R.K.; Koenigstein, M.L. Local bond strength of vitreous enamel coated rebar to concrete. Constr. Build. Mater. 2012, 35, 428–439.

13. Moser, R.D.; Allison, P.G.; Williams, B.A.; Weiss, C.A., Jr.; Diaz, A.J.; Gore, E.R.; Malone, P.G. Improvement in the geopolymer-to-steel bond using a reactive vitreous enamel coating. Constr. Build. Mater. 2013, 49, 62–69.

14. Chang, J.J.; Yeih, W.; Tsai, C.L. Enhancement of bond strength for epoxy-coated rebar using river sand. Constr. Build. Mater. 2002, 16, 465–472.

15. Chang, J.J.; Yeih, W. The effect of particle shape on bond strength improvement of epoxy-particle coating composites. J. Mater. Sci. Technol. 2001, 9, 153–160.

16. Yang, X.; Jha, A.; Brydon, R.; Cochrane, R.C. An analysis of the microstructure and interfacial chemistry of steel-enamel interface. Thin Solid Films 2003, 443, 33–45.

17. Samiee, L.; Sarpoolaky, H.; Mirhabibi, A. Microstructure and adherence of cobalt containing and cobalt free enamels to low carbon steel. Mater. Sci. Eng. A 2007, 458, 88–95.

18. Stanmore, B.R.; Gilot, P. Review-calcination and carbonation of limestone during thermal cycling for CO2 sequestration. Fuel Process. Technol. 2005, 86, 1707–1743.

19. Garcia-Labiano, F.; Abad, A.; de Diego, L.F.; Gayan, P.; Adanez, J. Calcination of calcium-based sorbent at pressure in a broad range of CO2 concentrations. Chem. Eng. Sci. 2002, 57, 2381–2393.

20. Jamil, H.E.; Shriri, A.; Boulif, R.; Montemor, M.F.; Ferreira, M.G.S. Corrosion behavior of reinforcing steel exposed to an amino alcohol based corrosion inhibitors. Cem. Concr. Compos. 2005, 27, 671–678.

21. Behzadnascab, M.; Mirabedini, S.M.; Kabiri, K.; Jamali, S. Corrosion performance of epoxy coatings containing silane treated ZrO2 nanoparticles on mild steel in 3.5% NaCl solution. Corros. Sci. 2011, 53, 89–98.

22. Jegdic, B.V.; Bajat, J.B.; Popic, J.P.; Stevanovic, S.J.; Miskovic-Stankovic, V.B. The EIS investigation of powder polymer coatings on phosphate low carbon steel: The effect of NaNO2 in the phosphating bath. Corros. Sci. 2011, 53, 2872–2880.

23. Rudd, A.L.; Breslin, C.B.; Mansfeld, F. The corrosion protection afforded by rare earth converstion coatings applied to magnesium. Corros. Sci. 2000, 42, 275–288.

24. Zhang, Y.; Shao, Y.; Zhang, T.; Meng, G.; Wang, F. High corrosion protection of a polyaniline/organophilic montmorillonite coating for magnesium alloy. Prog. Org. Coat. 2013, 76, 804–811.

25. Liu, Y.; Wang, J.; Liu, L.; Wang, F. Study of the failure mechanism of an epoxy coating system under high hydrostatic pressure. Corros. Sci. 2013, 74, 59–70.

26. Singh Raman, R.K.; Chakraborty Banerjee, P.; Lobo, D.E.; Gullapalli, H.; Sumandasa, M.; Kumar, A.; Choudhary, L.; Tkacz, R.; Ajayan, P.M.; Majumder, M. Protecting copper from electrochemical degradation by grapheme coating. Carbon 2012, 50, 4040–4045.

27. Hirschorn, B.; Orazem, M.E.; Tribollet, B.; Vivier, V.; Frateur, I.; Musiani, M. Determination of effective capacitance and film thickness form constant-phase-element parameters. Electrochim. Acta 2010, 55, 6218–6227.

28. Zhang, Y.; Shao, Y.; Zhang, T.; Meng, G.; Wang, F. The effect of epoxy coating containing emeraldine base and hydrofluoric acid doped polyaniline on the corrosion protection of AZ91D magnesium alloy. Corros. Sci. 2011, 53, 3747–3755.

29. Chakraborty Banerjee, P.; Singh Raman, R.K. Electrochemical impedance spectroscopic investigation of the role of alkaline pre-treatment in corrosion resistance of a silane coating on magnesium alloy ZE41. Electrochim. Acta 2011, 56, 3790–3798.

30. National Research Council (NRC). International Critical Tables; McGraw-Hill: Washington, DC, USA, 1927; Volume 2, p. 116.

CHAPTER 12

Effect of Annealing Temperature on the Mechanical and Corrosion Behavior of a Newly Developed Novel Lean Duplex Stainless Steel

Yanjun Guo [1], Jincheng Hu [2], Jin Li [1,*], Laizhu Jiang [2], Tianwei Liu [3] and Yanping Wu [4]

[1] Department of Materials Science, Fudan University, No. 220, Handan Road, Shanghai 200433, China

[2] Research and Development Center, Baosteel Co., Ltd., No. 885, Fujin Road, Shanghai 201900, China

[3] Science and Technology on Surface Physics and Chemistry Laboratory, P.O. Box 718-35, Mianyang 621907, Sichuan Province, China

[4] China Academy of Engineering Physics, P.O. Box 919-71, Mianyang 621900, Sichuan Province, China

ABSTRACT

The effect of annealing temperature (1000–1150 °C) on the microstructure evolution, mechanical properties, and pitting corrosion behavior of a newly developed novel lean duplex stainless steel with 20.53Cr-3.45Mn-2.08Ni-0.17N-0.31Mo was studied by means of optical metallographic microscopy (OMM), scanning electron microscopy (SEM), magnetic force microscopy (MFM), scanning Kelvin probe force microscopy (SKPFM), energy dispersive X-ray spectroscopy (EDS), uniaxial tensile tests (UTT), and potentiostatic critical pitting temperature (CPT). The results showed that tensile and yield strength, as well as the pitting corrosion resistance, could be degraded with annealing temperature increasing from 1000 up to 1150 °C. Meanwhile, the elongation at break reached the maximum of 52.7% after annealing at 1050 °C due to the effect of martensite transformation induced plasticity (TRIP). The localized pitting attack preferentially occurred at ferrite phase, indicating that the ferrite phase had inferior pitting corrosion resistance as compared to the austenite phase. With increasing annealing temperature, the pitting resistance equivalent number (PREN) of ferrite phase dropped, while that of the austenite phase rose. Additionally, it was found that ferrite possessed a lower Volta potential than austenite phase. Moreover, the Volta potential difference between ferrite and austenite increased with the annealing temperature, which was well consistent with the difference of PREN.

KEYWORDS

lean duplex stainless steel; annealing temperature; mechanical properties; TRIP effect; pitting corrosion behavior; PREN; Volta potential

1. INTRODUCTION

Duplex stainless steels (DSSs), strongly relying on a balanced two-phase microstructure of ferrite (α) and austenite (γ), have attractive mechanical and corrosion properties and are thus widely used in the chemical, petrochemical, nuclear, marine and paper industries [1,2,3,4]. In past years, the development of DSSs has followed two routes. On one side, great efforts have been made to improve the corrosion resistance of DSSs by increasing the content of chromium (Cr), molybdenum (Mo) and nitrogen (N), e.g., the DSS UNS S32750 with high mass fractions of 25%–27% Cr, 3%–4.5% Mo and 0.25%–0.28% N, which has been developed to meet with the requirement of good resistance and high strength in severe service environment [5,6,7]. On the other side, in order to conserve resources, "lean" route DSSs, e.g., UNS S32101 type with less nickel (Ni) but higher yield strength and better pitting resistance than standard austenitic grades, have been developed to meet the demand for grades with a lower cost [8,9,10,11,12]. However, the elongation of commercial UNS S32101 was around 30%, which limited the application of lean DSSs in many fields [8,13,14].

Recently, some developed lean DSSs with martensitic transformation have been designed to bring transformation induced plasticity (TRIP) effect and improve the ductility of lean DSSs. Herrera has designed a novel Mn based ductile lean duplex stainless TRIP steel (Fe-19.9Cr-0.42Ni-0.16N-4.79Mn-0.11C-0.46Cu-0.35Si, wt%), which had 1 GPa ultimate tensile strength and an elongation to fracture of above 60% resulted from a sequential martensite transformation of $\gamma \rightarrow \varepsilon \rightarrow \alpha'$ [14]. Choi has reported the effects of nitrogen addition on the strain-induced martensitic transformation of Fe-20Cr-5Mn-0.2Ni duplex stainless steel and found that the elongation was up to 60% in some lean DSSs containing 0.3% N [15]. In our work, a new lean duplex stainless steel with a composition of 20.53Cr-3.45Mn-2.08Ni-0.17N-0.31Mo has been developed. With a low Ni content, this resource-saving DSS can be less affected by price fluctuation of precious metal and reduce the risk of both production and uses. Additionally, the elongation has been improved to around 50% thanks to martensitic TRIP effect, which is obviously higher than that of 30% in S32101. Thus, this newly developed DSS holds a great potential to be the promising replacement of UNS S32101 in the advantage of improved properties.

In order to expand the application of this newly developed DSS used in this work, it is of great importance to obtain the suitable mechanical properties and good corrosion resistance. Solution heat treatment shows great impact to the mechanical and corrosion properties of DSS. The aim of the present work is to find the optimum solution heating temperature for the studied specimen based on investigating the relationship between the solution heating temperature and the corresponding mechanical properties and pitting corrosion resistance. The effect of annealing temperature on the microstructure, mechanical properties and corrosion behavior of this novel lean duplex stainless steel has been characterized by microscope, mechanical test and electrochemical technique.

2. EXPERIMENTAL SECTION

2.1. Materials

The studied material in the paper was a novel developed lean duplex stainless steel with the chemical composition shown in Table 1. It was melted in a 50 kg vacuum furnace and then cast as a single square ingot. After removing the oxide skin, the ingot was forged into square slabs at the temperature ranging from 900 to 1200 °C with a thickness of 40 mm. The slabs were reheated at 1200 °C for 2 h and hot-rolled, using a laboratory hot-rolling mill, into 4 mm thick plates and then cold-rolled into 1.5 mm. The specimens were machined into blocks of dimension of 10 mm × 10 mm. The phase diagram of the newly developed DSS used in present work was calculated by using the Thermo-calc software, as shown in Figure 1. It could be seen that mainly ferrite and austenite were formed in the temperature range between 1000 and 1200 °C. Moreover, the specimen aged between 500 and 980 °C was subjected to precipitation of secondary phases, such as Cr2N, σ and secondary austenite (γ2), which would seriously deteriorate the corrosion resistance of the duplex stainless steel [16,17]. Therefore, in the present work, the solution heating temperature below 1000 °C was not chosen so as to avoid the formation of these detrimental secondary phases. Thus, the specimens were subjected to different annealing treatment at 1000, 1020, 1050, 1080, 1110, and 1150 °C for 30 min, respectively, and quenched in water to avoid intermetallic precipitates.

Table 1. Chemical composition (wt%) of the duplex stainless steels (DSS).

UNS No.	C	Si	Mn	P	S	Cr	Ni	Mo	Cu	N
S32750	≤0.03	≤0.8	≤1.2	≤0.035	≤0.02	24.0/26.0	6.0/8.0	3.0/5.0	≤0.5	0.24/0.32
S32101	<0.04	<1.0	4.0/6.0	<0.04	<0.03	21.0/22.0	1.35/1.70	0.10/0.80	0.10/0.80	0.20/0.25
New DSS	0.03	0.32	3.45	0.01	0.004	20.53	2.08	0.31	0.34	0.17

2.2. Uniaxial Tensile Test

The uniaxial tensile tests were conducted at room temperature with specimens having a gage thickness of 1.5 mm and width of 20 mm annealed at different temperatures using an Instron 5985 (Instron company, Pittsburgh, PA, USA) testing machine according to BS EN ISO 6892-1:2009 [18]. The test was carried out at an extension rate of 3 mm/min before reaching yield point and 20 mm/min after reaching yield point. The test was conducted with three parallel specimens for every annealing temperature and the average data were used for determination of the tensile strength, yield strength and elongation at break values in this paper. In order to observe the microstructure of annealed specimens, which were pre-stretched to 40%, the specimens were etched in 2% HF solution for 30 s, to make the ferrite, austenite, and martensite be distinguished in the optical images.

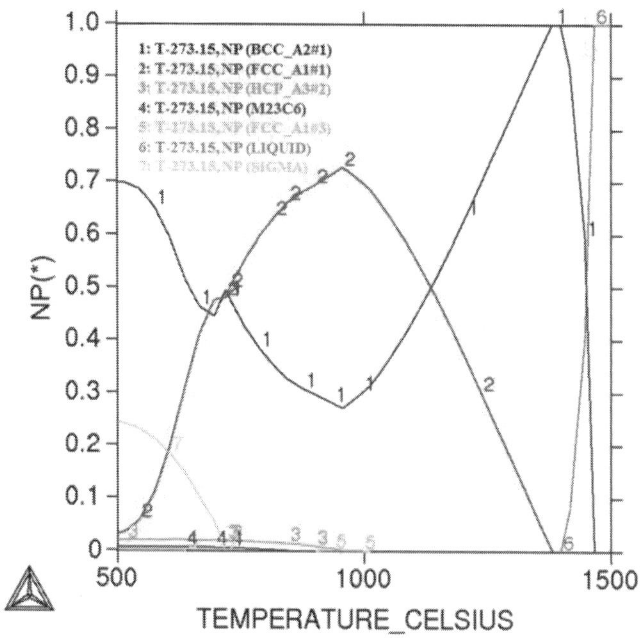

Figure 1. The phase diagram calculated by using the Thermo-Calc software for the newly developed DSS used in the present work.

2.3. Electrochemical Measurement

Electrochemical corrosion behavior of the specimens were performed with a CHI 660D potentiostat (Shanghai Chenhua

Instrument Co., Ltd, Shanghai, China). All the measurements were carried out with a three-electrode cell where a Pt foil acted as the counter electrode and a saturated calomel electrode (SCE) as the reference electrode. The specimens acting as working electrodes were sealed in epoxy resin. Prior to each experiment, the working electrode was wet mechanically ground, subsequently polished with a 1.5 μm diamond paste, ultrasonically cleaned in ethanol and distilled water, and then dried in air. To avoid the crevice corrosion, interfaces between specimen and resin were sealed with silica gel sealant, and dried in air. The exposed electrode surface area was 100 mm2. The test solution was 500 mL of 1 mol/L NaCl. According to ASTM G 150-99 (2010) [19], critical pitting temperature (CPT) determination of this novel lean duplex stainless steel was performed via potentiostatic measurements by applying a constant potential of 250 mV SCE and increasing the solution temperature at a rate of 1 °C/min from 2 °C. The CPT was defined as the temperature at which the current sharply increased to 100 μA. After testing, specimens were electrochemically etched for 8 s in 30 wt% KOH solution at applied voltage of 2 V, to make the ferrite and austenite be distinguished in the optical images, and then examined for pits [20]. The CPT measurements of the same specimen for every annealing temperature were repeated at least six times, and the typical data were chosen as the CPT values in this work.

2.4. Optical Metallographic Microscopy and SEM-EDS Analysis

Microstructural examination of specimens after the electrochemical measurements was conducted using optical metallographic microscopy and SEM (SCE Phillips XL30 FEG, Amsterdam, The Netherlands). The volume fraction of ferrite and austenite was evaluated by quantitative metallography based on the software attached to the microscope. Energy dispersive X-ray spectroscopy (EDS) was performed on the individual phases to obtain their elemental composition. Because of the insensitivity of EDS techniques to N, an approximate calculation was conducted to determine the N fraction in both phases. The level of N in ferrite phase was assumed to be the saturation value (around 0.05%) and that of austenite was calculated based on the content of N in the whole alloy and the phase volume fraction [21]. The SEM-EDS analysis was tested with six parallel specimens for every annealing temperature. Each value of the element distribution in Table 2 was an average of more than ten measurements.

Table 2. Chemical composition and pitting resistance equivalent number (PREN) of the two phases for specimens annealed at different temperatures, obtained by EDS.

Annealing temperature	Phase	Volume fraction (%)	Cr (%)	Ni (%)	Mo (%)	N (%)	$PREN_{16}$	$PREN_{30}$
1000 °C	Ferrite	52.1	21.59	1.51	0.40	0.05	23.710	24.410
	Austenite	47.9	19.37	2.70	0.21	0.30	24.863	29.063
1020 °C	Ferrite	57.9	21.45	1.68	0.32	0.05	23.306	24.006
	Austenite	42.1	19.27	2.62	0.29	0.34	25.667	30.427
1050 °C	Ferrite	61.4	21.24	1.72	0.33	0.05	23.129	23.829
	Austenite	38.6	19.40	2.66	0.28	0.36	26.084	31.124
1080 °C	Ferrite	62.9	21.25	1.52	0.32	0.05	23.106	23.806
	Austenite	37.1	19.31	3.03	0.30	0.37	26.220	31.400
1110 °C	Ferrite	66.2	21.15	1.73	0.35	0.05	23.105	23.805
	Austenite	33.8	19.32	2.77	0.22	0.41	26.606	32.346
1150 °C	Ferrite	69.1	21.05	1.85	0.32	0.05	22.906	23.606
	Austenite	30.9	19.37	2.60	0.29	0.44	27.367	33.527

2.5. Magnetic Force Microscopy (MFM) and Scanning Kelvin Probe Force Microscopy (SKPFM) Measurements

The MFM and SKPFM measurements were performed using a dimension Nanoscope V (Bruker Corporation, Santa Barbara, CA,

USA). All measurements were conducted in air at room temperature and an ambient relative humidity of about 45%. The probes used in MFM measurements were Nanoprobe™ (Bruker Corporation, Santa Barbara, CA, USA) MSEP Co/Cr-coated tips with force constant of 2.8 N/m and resonant frequency of 75 kHz, while the probes in SKPFM measurements were SCM-PIT conductive Pt/Ir-coated tips with the same force constant and resonant frequency. A dual-scan mode was used to record a second signal in addition to surface topography signal in MFM and SKPFM measurements. The topography of the sample surface was obtained in the first scan with the tapping mode. In the second scan, the cantilever was lifted up 100 nm to avoid the influence of topographic features. The principle and details of the SKPFM measurement have been reported previously [22,23].

3. RESULTS AND DISCUSSION

3.1. Microstructures and Elements Distribution

Figure 2 showed the microstructure of specimens annealed at different temperatures for 30 min. The island-shaped austenite phase was evenly embedded in the band-shaped ferrite matrix. Figure 3 displayed the volume fraction of the ferrite/austenite phases against annealing temperature. It could be observed that as the annealing temperature ranged from 1000 to 1150 °C, the ferrite region gradually increased and austenite region decreased, which was a typical diffusion-controlled solid-state phase transformation: $\gamma \rightarrow \alpha$ and determined by the equilibrium phase diagram [24]. Besides, the ferrite and austenite phase approximately reached balanced when annealed at 1000 °C (i.e., VolumeFerrite:VolumeAustenite ≈ 1:1). However, Figure 1 revealed that the equal volume fraction of the two phases in the investigated specimens was achieved at annealing temperature higher than 1000 °C, which was attributed to the fact that the Thermo-Calc described thermodynamic equilibrium conditions while the steels could never be in true equilibrium [10]. The content of main alloying elements measured by EDX in austenite and ferrite phases was listed in Table 2. It was shown that alloying elements, such as Cr and Mo, were enriched in the ferrite phase, while N and Ni were concentrated in the austenite phase. As the annealing temperature was increased, content of Cr and Mo decreased in the ferrite phase, while the content of Cr and Mo in austenite increased slightly. It could be explained by that the higher annealing temperature, resulted in the higher volume fraction of ferrite, which further led that corresponding Cr and Mo were spread over a larger region and diluted [20].

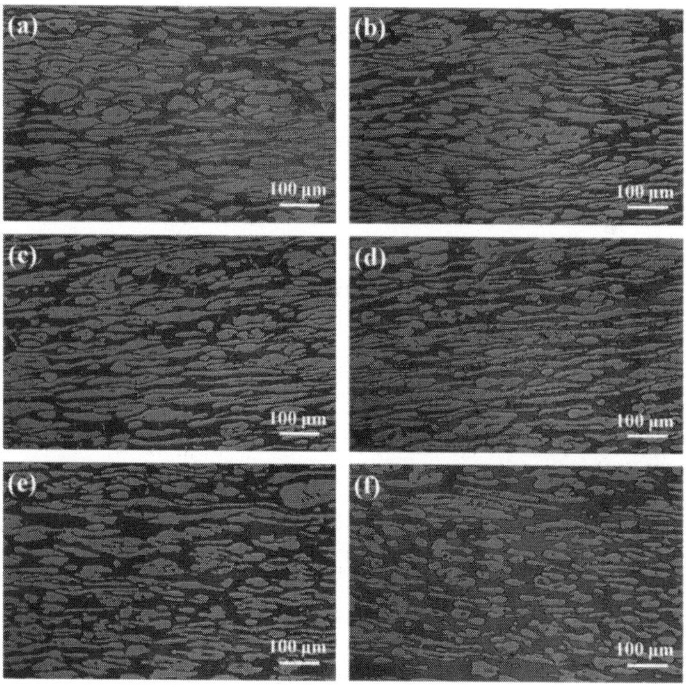

Figure 2. Optical micrographs of specimens annealed at different temperature for 30 min, the dark area representing ferrite while the light representing austenite. (a) 1000 °C; (b) 1020 °C; (c) 1050 °C; (d) 1080 °C; (e) 1110 °C; (f) 1150 °C.

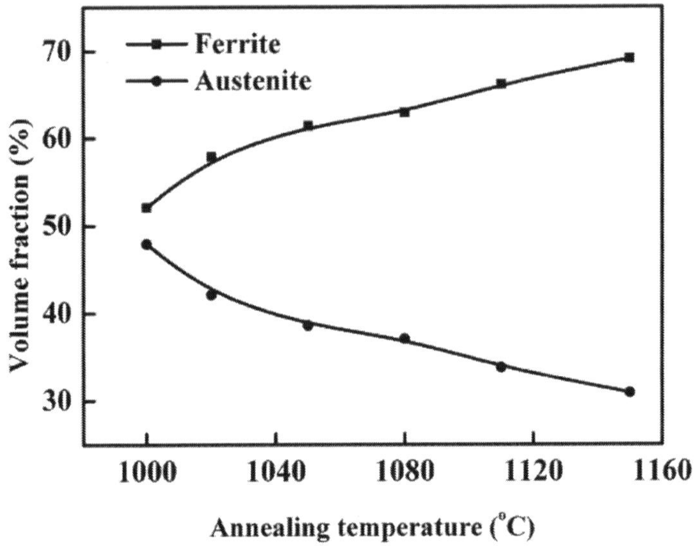

Figure 3. Volume fraction of ferrite and austenite phase with different annealing temperatures for 30 min.

3.2. Mechanical Properties

Figure 4 and Figure 5, respectively, presented the tensile and yield strength of this novel lean duplex stainless steel as function of annealing temperature. The results showed that the tensile strength and yield strength decreased with increasing the annealing temperature from 1000 to 1150 °C. It was revealed that at 1000 °C, the specimens owned the highest tensile and yield strength, which could be attributed to the fact that the tensile and yield strength depended strongly on grain size [2,25,26]. According to the well-known Hall-Petch relation, a smaller average grain size resulted in a stronger material [27]. As the annealing temperature was increased from 1000 up to 1150 °C, the grain growth could be enhanced leading to the increased grain size, thus, the tensile and yield strength decreased. In addition, additional strengthening in DSSs could be caused by solid solution hardening [2]. The solid solution strengthening effects of chromium, molybdenum and nitrogen reached the maximum at 1000 °C as shown by the increased additions of alloying elements in Table 2. Figure 6 showed the elongation at break of this novel lean duplex stainless steel as function of annealing temperature. The elongation at break reached the maximum of 52.7% after annealing at 1050 °C and the minimum of 47.3% at annealing temperature of 1150 °C, being obviously higher than that of 25% in S32750 and 30% in S32101, which might be attributed to the martensite transformation [1,8,13,14]. Typical values of mechanical properties of solution annealed DSS were summarized in Table 3. The microstructure of specimen annealed at 1050 °C and pre-stretched to 40% is shown in Figure 7. It could be found that much lath-shaped martensite was embedded in dark retained austenite phase, which exhibited that austenite gradually transformed into martensite during tensile deformation. Moreover, it was already demonstrated that the strain-induced martensite transformation was helpful for improving the plasticity of the lean DSS associated with TRIP effect [15]. Meantime, it has been reported that the martensite transformation of lean DSS proceeded from austenite to hexagonal martensite to near cubic martensite ($\gamma \rightarrow \varepsilon \rightarrow \alpha'$) [13,14]. However, Zhang has indicated that the main reason for the improvement of elongation of lean DSS was the $\gamma \rightarrow \varepsilon$ transformation, not in proportion to the generation of the α' martensite [14]. For this novel lean DSS, the martensite transformation process remained to be investigated in detail. The elongation at break of specimen annealed at 1000 °C reached 51.13%, which was slightly lower than that of specimen after annealing at 1050 °C. However, combining the tensile and yield strength, the specimen annealed at 1000 °C demonstrated the best mechanical properties.

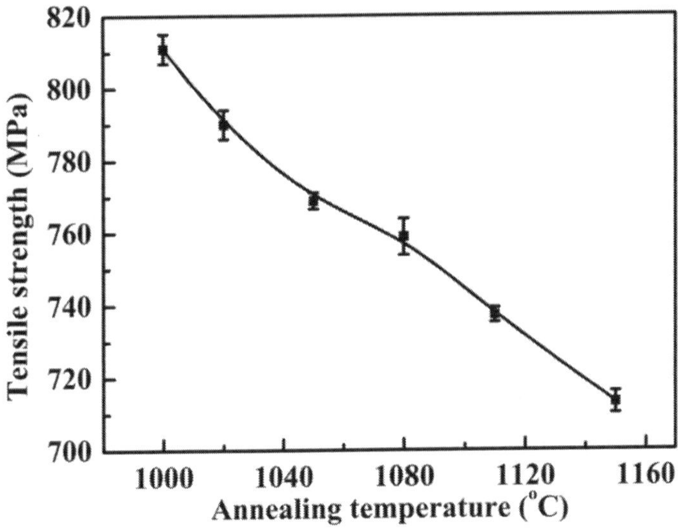

Figure 4. Tensile strength of specimens with different annealing temperatures for 30 min.

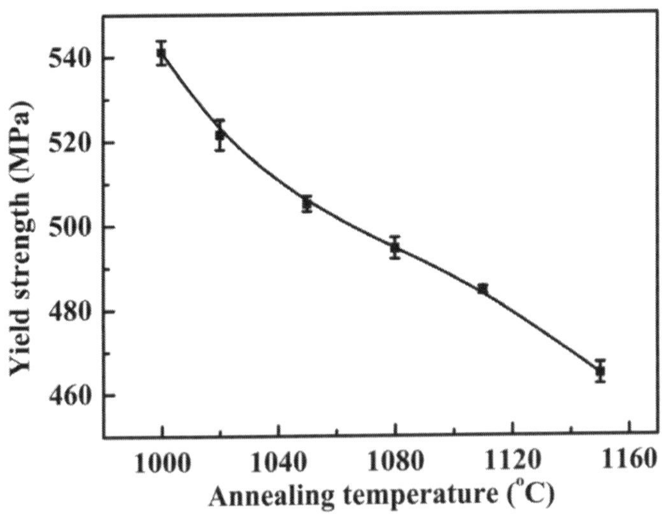

Figure 5. Yield strength of specimens with different annealing temperatures for 30 min.

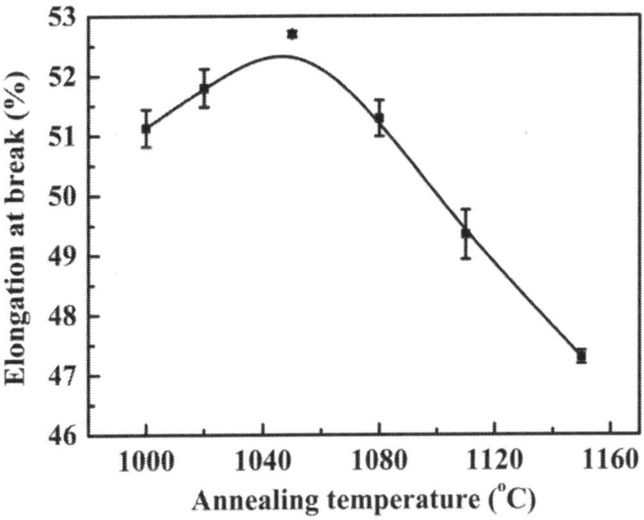

Figure 6. Elongation at break of specimens with different annealing temperatures for 30 min.

Table 3. Typical values of mechanical properties of solution annealed duplex stainless steels (DSS).

Alloy	Ultimate tensile strength, MPa	0.2% proof stress (min) MPa	Elongation (min) A5, %
UNS S32750 [1]	800–1000	550	25
UNS S32101 [8]	650–700	450	30
Newly Developed DSS	737–811	465	47.3

3.3. Pitting Corrosion Behavior

Figure 8 illustrated the typical curves of current density against annealing temperature for specimens annealed at different temperatures for 30 min. The current density presented a lower value less than 1 μA/cm2 during the initial heating, indicating that the steel surface was protected by the passive film. With the solution temperature increasing to CPT, some current density fluctuations were found, which were associated with a passive film breakdown in the form of metastable pits. As the temperature increased further reaching to the CPT, the current density rose sharply due to the occurrence of the stable pits. As the annealing temperature increased from 1000 to 1150 °C, the CPT decreased from 40.8 to 27 °C. The highest CPT, 40.8 °C, was obtained when the specimens were annealed at 1000 °C, which meant that the specimens annealed at 1000 °C showed the best pitting corrosion resistance. Compared with the lean duplex stainless steel UNS S32101, of which the highest CPT of 35.2 °C was obtained at annealing temperature of 1000 °C, this novel lean duplex stainless steel showed a better pitting corrosion resistance [10]. This could be explained by that this novel steel had higher content of N and lower content of Manganese (Mn) than S32101 [28].

Figure 7. Optical micrographs of specimen annealed at 1050 °C for 30 min and pre-stretched to 40%, the dark area representing austenite while the light representing ferrite, and lath-shaped martensite was embedded in austenite phase.

The morphologies of specimens were observed under SEM after electrochemical test and stable pits could be found. The stable pitting of specimens annealed at 1000 °C was located at both ferrite phase and austenite phase, while those of specimens annealed at temperatures over 1000 °C mainly distributed from both two phases to ferrite phase and the pitting corrosion occurred randomly. Especially for specimens annealed at 1150 °C, the stable pitting sites mainly attributed in the ferrite phase, and the pitting corrosion preferentially occurred in the ferrite phase. Typical stable pitting morphologies of the specimens annealed at 1000 °C, 1050 °C, and 1150 °C are shown in Figure 9.

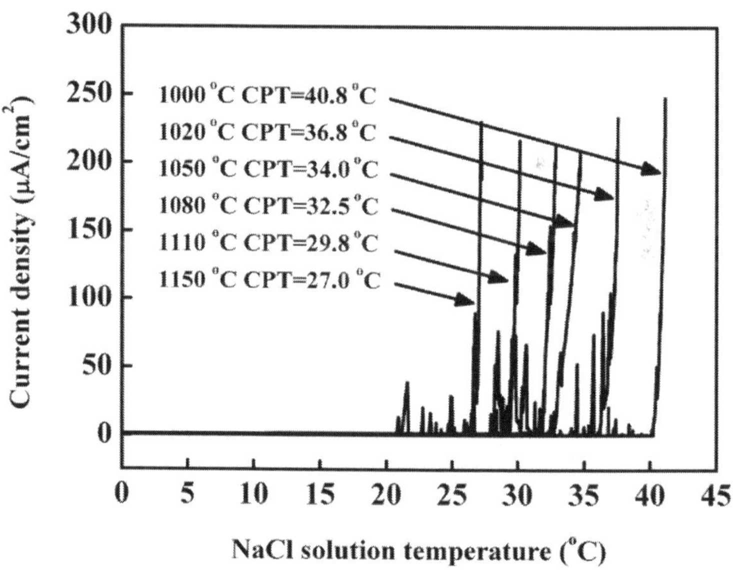

Figure 8. Curves of current density and temperature for specimens annealed at different temperatures in 1 M NaCl solution at an applied potential of 250 mV (SCE).

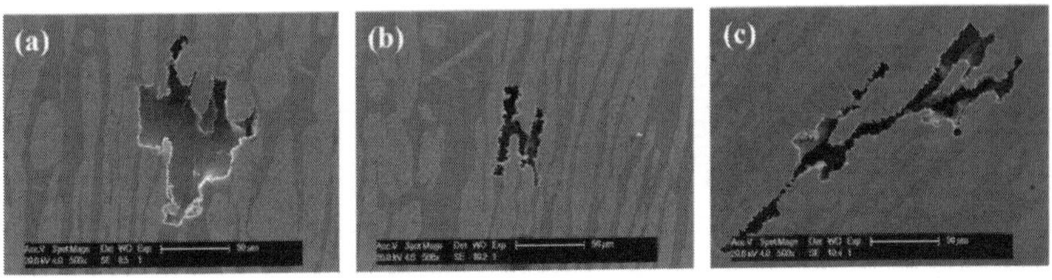

Figure 9. SEM morphologies of pittings formed on specimens annealed at different temperature for 30 min after the CPT test. (a) 1000 °C; (b) 1050 °C; (c) 1150 °C.

The resistance to pitting corrosion in duplex stainless steels was found to be strongly dependent on the chemical composition. It was known that the pitting resistance equivalent number of single phase could be used to evaluate the corrosion resistance of duplex stainless steels quickly. The empirical formula of pitting resistance equivalent number was defined by the following equation:

$$PREN_{16\text{–}30} = \text{wt\% Cr} + 3.3\ \text{wt\% Mo} + 16\text{–}30\ \text{wt\% N} \quad (1)$$

Different subscript varying in this formula indicated the different nitrogen coefficient [28,29,30]. The alloy (or phase) with higher pitting resistance equivalent number would exhibit better pitting corrosion resistance. Both the PREN16 and PREN30 of ferrite and austenite were calculated and listed in Table 2. Based on this formula, as the annealing temperature increased, the pitting resistance equivalent number of ferrite phase decreased while that of austenite phase increased. Figure 10 showed the pitting resistance equivalent number of single phase as function of annealing temperature. Such results were caused by the change of chemical composition of each phase corresponding to the variation of annealing temperature [31,32,33]. It has been reported that the pitting resistance of duplex stainless steels was determined by the weaker phase not the entire alloy [7,34]. For this novel lean duplex stainless steel, when the specimens were annealed at the temperature 1000 °C, both PREN16 and PREN30 differences between austenite and ferrite reached the lowest, which meant the two phases had similar pitting resistance at 1000 °C. Therefore, the pitting was propagated at both phases, shown in Figure 9a. However, the difference of PREN16 between two phases was lower than that of PREN30, indicating that it was better to use PREN16 to explain the pitting corrosion behavior, which was in good agreement with the reported literature: 16 was the most frequently used subscript [2]. When the annealing temperature exceeded 1000 °C, the pitting resistance equivalent number of austenite phase was much higher than that of ferrite phase, and pitting mainly occurred in the ferrite phase. The pitting morphologies of specimen annealed at 1150 °C showed that pits were located mainly at ferrite areas, as shown in Figure 9c, indicating that the ferrite phase was much weaker than austenite phase. It could be found that the stable pitting morphologies of the specimens after CPT tests were fairly consistent with the pitting resistance equivalent number of the weaker phase for this new lean duplex stainless steel. It also could be obtained that the CPT reflected the pitting corrosion resistance of the weaker phase and that the pitting corrosion resistance of duplex stainless was determined by the weaker phase not the entire alloy.

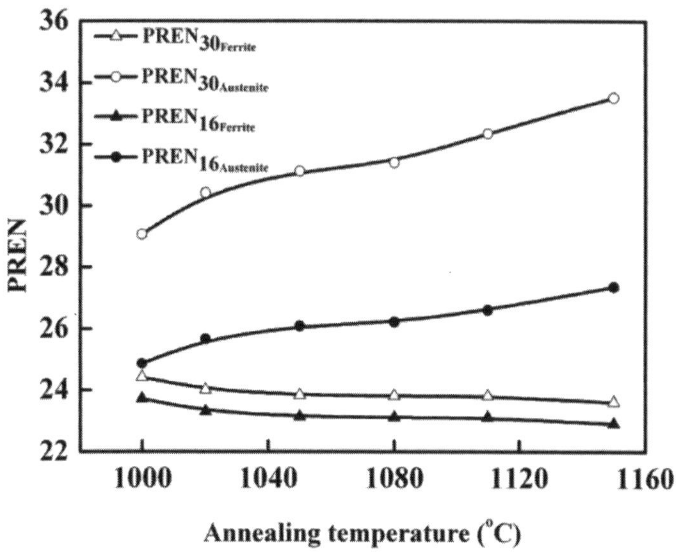

Figure 10. PREN for each phase of specimens with different annealing temperatures for 30 min.

3.4. MFM and SKPFM Results

Figure 11 showed the microstructure of specimens after annealing treatment at 1000, 1050 and 1150 °C for 30 min using MFM and SKPFM. Figure 11a1–a3 respectively displayed the MFM images of the specimens annealed at 1000, 1050 and 1150 °C with 80 μm × 80 μm, where striped ferromagnetic ferrite domain embedded in the paramagnetic austenite. Figure 11b1–b3 respectively revealed the SKPFM images of the same area with 80 μm × 80 μm. Figure 11c1–c3 demonstrated the Volta potential differences between the ferrite and austenite of specimens. The ferromagnetic ferrite had a lower Volta potential than the austenite. This was the same propensity as seen in the previous work [35,36,37]. The Volta potential difference between ferrite and austenite increased as the increase of annealing temperature. As shown in Table 4, the average potential difference between ferrite and austenite for the specimens annealed at 1000 °C, 1050 °C, 1150 °C were about 37.5 mV, 50 mV, and 75 mV. The highest value of Volta potential difference was obtained from specimen annealed at 1150 °C, and the lowest was from the specimen annealed at 1000 °C. A linear relationship between the Volta potential measured by SKPFM and corrosion potential has been reported [38,39,40,41,42]. It was reported that the possible galvanic interactions between the ferrite and austenite phases had been obtained by SKPFM technique [43]. In this paper, the ferrite had a lower Volta potential in this novel lean duplex stainless steel using SKPFM, which was in good agreement with the results in potentiostatic measurement (CPT) that pitting corrosion was found to be preferentially occurred in ferrite phase and the ferrite was a weaker phase. Compared with SKPFM technique, CPT measurement was performed at an applied potential. It couldn't be explained by galvanic interactions. However, a conclusion might be obtained that the pitting corrosion tendency of ferrite increased when applying a static potential. In addition, the Volta potential difference between ferrite and austenite was well consistent with the difference between PRENFerrite and PRENAustenite, which could be explained by that the elements of Cr, Mo, and N should have certain influences on the Volta potential [35].

4. CONCLUSIONS

A novel lean duplex stainless steel 20Cr-2Ni-3Mn-0.17N-0.31Mo has been developed with higher elongation at break and better pitting corrosion resistance than UNS S32101. It has a great potential to be the promising replacement of S32101 in the advantage of improved properties. In order to expand the application of this new developed duplex steel, it is of great importance to obtain the suitable mechanical properties and good corrosion resistance. Thus, the effect of annealing temperature on the microstructure, mechanical properties and corrosion behavior of this novel lean duplex stainless steel has been characterized by microscope, mechanical test and electrochemical technique. Based on the experimental results, it is discovered that the best mechanical properties and pitting corrosion resistance of this newly developed DSS can be obtained at annealing temperature 1000 °C, which should be of great help in the industrial production. The main conclusions are as follows.

(1) The volume fraction of austenite decreased continuously and the ferrite increased with the annealing temperature increasing from 1000 to 1150 °C.

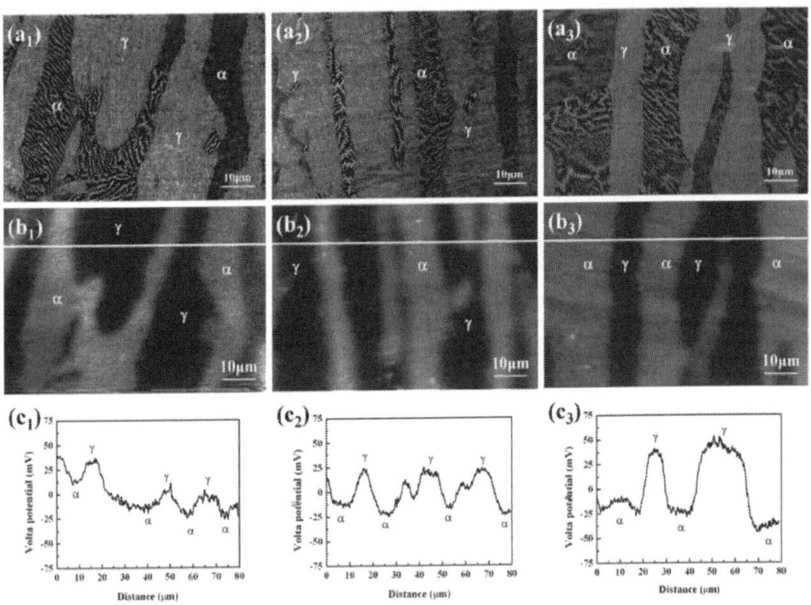

Figure 11. (a1–a3) MFM and (b1–b3) SKPFM images of specimens respectively annealed at 1000 °C, 1050 °C and 1150 °C for 30 min; (c1–c3) volta potential difference profile of a line as showed in (b1–b3).

Table 4. The Volta potential difference between ferrite and austenite phase for the specimens annealed at 1000, 1050, and 1150°C.

Annealing temperature (°C)	Average Volta potential difference$_{Ferrite\text{-}Austenite}$ (mV)
1000	37.5
1050	50
1150	75

(2) The tensile strength and yield strength decreased with the increase of annealing temperature, which could be explained by the grain size and solid solution strengthening effects of alloying elements.

(3) The elongation at break reached the maximum of 52.7% after annealing at 1050 °C due to martensite transformation associated with TRIPeffect.

(4) The critical pitting temperature decreased with increasing of annealing temperature. The localized pitting attack preferentially occurred at ferrite phase. The pitting corrosion behaviorin chloride solutions could be explained by redistribution of main alloying elements, such as chromium, molybdenum and nitrogen in two phases, resulting in the change of PREN16 of ferrite and austenite phases.

(5) Ferrite had a lower Volta potential than austenite phase. The Volta potential difference between ferrite and austenite increased as the increase of annealing temperature. The Volta potential difference between ferrite and austenite was in good conformity with the difference between PRENFerrite and PRENAustenite.

ACKNOWLEDGMENTS

The financial support from the National Natural Science Fund of China (Grants No. 51371053, 51131008 and 51134010), the National Key Technology R&D Program (2012BAE04B00), and the Discipline Development Fund of Science and Technology on Surface Physics and Chemistry Laboratory (ZDXKFZ201213) are gratefully acknowledged.

AUTHOR CONTRIBUTIONS

The specimens were supported by Laizhu Jiang. The experiments were carried out by Yanjun Guo and Jincheng Hu, morphology characterization was performed by Tianwei Liu and Yanping Wu, the experimental scheme was framed by Jin Li. The manuscript was composed by Yanjun Guo and revised by Jincheng Hu.

REFERENCES

1. Nilsson, J.O. Super duplex stainless-steels. Mater. Sci. Technol. 1992, 8, 685–700.
2. Chen, T.H.; Yang, J.R. Effects of solution treatment and continuous cooling on sigma-phase precipitation in a 2205 duplex stainless steel. Mater. Sci. Eng. A 2001, 311, 28–41.
3. He, H.; Zhang, T.; Zhao, C.Z.; Hou, K.; Meng, G.Z.; Shao, Y.W.; Wang, F.H. Effect of alternating voltage passivation on the corrosion resistance of duplex stainless steel. J. Appl. Electrochem. 2009, 39, 737–745.

4. Lee, K.M.; Cho, H.S.; Choi, D.C. Effect of isothermal treatment of SAF 2205 duplex stainless steel on migration of delta/gamma interface boundary and growth of austenite. J. Alloys Compd. 1999, 285, 156–161.

5. Nilsson, J.O.; Wilson, A. Influence of isothermal phase-transformations on toughness and pitting corrosion of super duplex stainless-steel SAF-2507. Mater. Sci. Technol. 1993, 9, 545–554.

6. Sato, Y.S.; Nelson, T.W.; Sterling, C.J.; Steel, R.J.; Pettersson, C.O. Microstructure and mechanical properties of friction stir welded SAF 2507 super duplex stainless steel. Mater. Sci. Eng. A 2005, 397, 376–384.

7. Weber, L.; Uggowitzer, P.J. Partitioning of chromium and molybdenum in super duplex stainless steels with respect to nitrogen and nickel content. Mater. Sci. Eng. A 1998, 242, 222–229.

8. Alfonsson, E. Lean Duplex—The First Decade of Service Experience. In Proceedings of the 8th Duplex Stainless Steels Conference, Beaune, France, 13–15 October 2010.

9. Zhang, W.; Jiang, L.Z.; Hu, J.C.; Song, H.M. Effect of ageing on precipitation and impact energy of 2101 economical duplex stainless steel. Mater. Charact. 2009, 60, 50–55.

10. Zhang, L.H.; Zhang, W.; Jiang, Y.M.; Deng, B.; Sun, D.M.; Li, J. Influence of annealing treatment on the corrosion resistance of lean duplex stainless steel 2101. Electrochim. Acta 2009, 54, 5387–5392.

11. Sieurin, H.; Sandstrom, R.; Westin, E.M. Fracture toughness of the lean duplex stainless steel LDX 2101. Metall. Mater. Trans. A 2006, 37A, 2975–2981.

12. Zhang, L.H.; Jiang, Y.M.; Deng, B.; Zhang, W.; Xu, J.L.; Li, J. Effect of aging on the corrosion resistance of 2101 lean duplex stainless steel. Mater. Charact. 2009, 60, 1522–1528.

13. Zhang, W.; Hu, J.H. Effect of annealing temperature on transformation induced plasticity effect of a lean duplex stainless steel. Mater. Charact. 2013, 79, 37–42.

14. Herrera, C.; Ponge, D.; Raabe, D. Design of a novel Mn-based 1 GPa duplex stainless TRIP steel with 60% ductility by a reduction of austenite stability. Acta Mater. 2011, 59, 4653–4664.

15. Choi, J.Y.; Ji, J.H.; Hwang, S.W.; Park, K.T. Strain induced martensitic transformation of Fe-20Cr-5Mn-0.2Ni duplex stainless steel during cold rolling: Effects of nitrogen addition. Mater. Sci. Eng. A 2011, 528, 6012–6019.

16. Pohl, M.; Storz, O.; Glogowski, T. Effect of intermetallic precipitations on the properties of duplex stainless steel. Mater. Charact. 2007, 58, 65–71.

17. Deng, B.; Wang, Z.Y.; Jiang, Y.M.; Sun, T.; Xu, J.L.; Li, J. Effect of thermal cycles on the corrosion and mechanical properties of UNS S31803 duplex stainless steel. Corros. Sci. 2009, 51, 2969–2975.

18. British Standards (BS) European Norm (EN) International Organization for Standardization (ISO) 6892-1:2009, "Metallic materials Tensile testing; Part 1: Method of test at ambient temperature". European Committee for Standardization: Brüssel, 2009.

19. American Society for Testing and Materials (ASTM) Standard G150-99. In Standard Test Method for Electrochemical Critical Pitting Temperature Testing of Stainless Steels, Annual Book of ASTM Standards; ASTM International: West Conshohocken, PA, USA, 2010.

20. Tan, H.; Jiang, Y.M.; Deng, B.; Sun, T.; Xu, J.L.; Li, J. Effect of annealing temperature on the pitting corrosion resistance of super duplex stainless steel UNS S32750. Mater. Charact. 2009, 60, 1049–1054.

21. GarfiasMesias, L.F.; Sykes, J.M.; Tuck, C.D.S. The effect of phase compositions on the pitting corrosion of 25 Cr duplex stainless steel in chloride solutions. Corros. Sci. 1996, 38, 1319–1330.

22. Nonnenmacher, M.; Oboyle, M.P.; Wickramasinghe, H.K. Kelvin probe force microscopy. Appl. Phys. Lett. 1991, 58, 2921–2923.

23. Rohwerder, M.; Turcu, F. High-resolution Kelvin probe microscopy in corrosion science: Scanning Kelvin probe force microscopy (SKPFM) versus classical scanning Kelvin probe (SKP). Electrochim. Acta 2007, 53, 290–299.

24. Yousefieh, M.; Shamanian, M.; Saatchi, A. Influence of step annealing temperature on the microstructure and pitting corrosion resistance of SDSS UNS S32760 welds. J. Mater. Eng. Perform. 2011, 20, 1678–1683.

25. Offerman, S.E.; van Dijk, N.H.; Sietsma, J.; Grigull, S.; Lauridsen, E.M.; Margulies, L.; Poulsen, H.F.; Rekveldt, M.T.; van der Zwaag, S. Grain nucleation and growth during phase transformations. Science 2002, 298, 1003–1005.

26. Badji, R.; Bacroix, B.; Bouabdallah, M. Texture, microstructure and anisotropic properties in annealed 2205 duplex stainless steel welds. Mater. Charact. 2011, 62, 833–843.

27. Zhang, X.D.; Hansen, N.; Gao, Y.K.; Huang, X.X. Hall-Petch and dislocation strengthening in graded nanostructured steel. Acta Mater. 2012, 60, 5933–5943.

28. Merello, R.; Botana, F.J.; Botella, J.; Matres, M.V.; Marcos, M. Influence of chemical composition on the pitting corrosion

resistance of non-standard low-Ni high-Mn-N duplex stainless steels. Corros. Sci. 2003, 45, 909–921.

29. Bae, S.H.; Lee, H.W. Effect of Mo contents on corrosion behaviors of welded duplex stainless steel. Met. Mater. Int. 2013, 19, 563–569.

30. Munoz, A.I.; Anton, J.G.; Nuevalos, S.L.; Guinon, J.L.; Herranz, V.N. Corrosion studies of austenitic and duplex stainless steels in aqueous lithium bromide solution at different temperatures. Corros. Sci. 2004, 46, 2955–2974.

31. Zhong, C.; Liu, F.; Wu, Y.T.; Le, J.J.; Liu, L.; He, M.F.; Zhu, J.C.; Hu, W.B. Protective diffusion coatings on magnesium alloys: A review of recent developments. J. Alloys Compd. 2012, 520, 11–21.

32. Zhong, C.; He, M.F.; Liu, L.; Wu, Y.T.; Chen, Y.J.; Deng, Y.D.; Shen, B.; Hu, W.B. Lower temperature fabrication of continuous intermetallic coatings on AZ91D magnesium alloy in molten salts. J. Alloys Compd. 2010, 504, 377–381.

33. Le, J.J.; Liu, L.; Liu, F.; Deng, Y.D.; Zhong, C.; Hu, W.B. Interdiffusion kinetics of the intermetallic coatings on AZ91D magnesium alloy formed in molten salts at lower temperatures. J. Alloys Compd. 2014, 610, 173–179.

34. Mori, G.; Bauernfeind, D. Pitting and crevice corrosion of superaustenitic stainless steels. Mater. Corros. 2004, 55, 164–173. [Google Scholar] [CrossRef]

35. Sathirachinda, N.; Pettersson, R.; Pan, J.S. Depletion effects at phase boundaries in 2205 duplex stainless steel characterized with SKPFM and TEM/EDS. Corros. Sci. 2009, 51, 1850–1860.

36. Guo, L.Q.; Li, M.; Shi, X.L.; Yan, Y.; Li, X.Y.; Qiao, L.J. Effect of annealing temperature on the corrosion behavior of duplex

stainless steel studied by in situ techniques. Corros. Sci. 2011, 53, 3733–3741.

37. Sathirachinda, N.; Gubner, R.; Pan, J.; Kivisakk, U. Characterization of phases in duplex stainless steel by magnetic force microscopy/scanning kelvin probe force microscopy. Electrochem. Solid State Lett. 2008, 11, C41–C45.

38. Stratmann, M. The investigation of the corrosion propereties of metals, covered with adsobed electrolyte layers—A new experimental technique. Corros. Sci. 1987, 27, 869–872.

39. Stratmann, M.; Streckel, H. On the atmospheric corrosion of metals which are covered with thin electrolyte layers.1. Verification of the experimental technique. Corros. Sci. 1990, 30, 681–696.

40. Yee, S.G.; Oriani, R.A.; Stratmann, M. Application of a kelvin microprobe to the corrosion of metals in humid atmospheres. J. Electrochem. Soc. 1991, 138, 55–61.

41. Schmutz, P.; Frankel, G.S. Characterization of AA2024-T3 by scanning Kelvin probe force microscopy. J. Electrochem. Soc. 1998, 145, 2285–2295.

42. Guillaumin, V.; Schmutz, P.; Frankel, G.S. Characterization of corrosion interfaces by the scanning Kelvin probe force microscopy technique. J. Electrochem. Soc. 2001, 148, B163–B173.

43. Femenia, M.; Canalias, C.; Pan, J.; Leygraf, C. Scanning Kelvin probe force microscopy and magnetic force microscopy for characterization of duplex stainless steels. J. Electrochem. Soc. 2003, 150, B274–B281.

Index